Small Steps Towards Abundance

Crops for a More Sustainable Agriculture

Front cover photos by Ruti Rubenstein
Author photo by William Cadman

The Thistle Syndicate is a group of writers and researchers who are dedicated to bringing specific skills and information to the public in a series of low cost publications.

ISBN 0-9741190-1-6

Biblio Books

Biblio Books Israel, Acco ○ Biblio Books International, Miami
http://www.bibliobooks.com ○email info@bibliobooks.com

To my mother Gerry who gave me a magnifying glass when I was very young and sent me out to look at the world.

Table of Contents

List of Colorplates

List of Figures

Small Steps Towards Abundance

Crops for a More Sustainable Agriculture

I. The Supermarket's False Abundance

The shops and supermarkets of the developed world paint a false picture of abundance. Acres of shelves contain every prepared food imaginable. There are thousands of square miles of glistening perfect to the eye fruits and vegetables and endless displays of refrigerated meat. Why then should we worry about food?

Actually these sights are like a painted screen that hides a desolate landscape.

The abundance of the supermarket is a false one.

In truth much of humanity is chronically, mercilessly hungry. The World Health Organization (WHO) estimate at the beginning of the year 2000 put the number of the world's hungry at one and a half billion (Gardner and Halweil, 2000, pg. 25).

In truth many citizens of the developed world are simultaneously overfed and undernourished, reaping the consequences in obesity, early death and general ill health.

In truth the food supply of the world is precarious and vulnerable, balanced as if on the blade of a knife.

True abundance would be a situation in which sufficient quantities of high quality, safe and nourishing food would be available for every citizen of the planet. This would be accomplished without destroying the soil and water, wasting non-renewable resources or wiping other living creatures off the face of the earth.

Is it Possible to Create Abundance in the World?

I believe that it is. But to reach true abundance it is first important to identify the problems and the challenges that face us now.

Quality and Purity One of the most troubling aspects of the food problem is the quality of food available.

In the age of poisoned grapes, irradiated apples and frozen pizza which contains no tomatoes and no cheese, many consumers wonder if the food we buy in the supermarkets has any nutritional value at all. Others worry about pesticide and herbicide residue and irradiation methods, which extend shelf life but radically change the food at a molecular level. Another concern is the safety of prepared foods that seem to have been whipped up in a test tube. Many people would like to know exactly what those letters and numbers mean. They also want to know why we have to eat butylated hydroxytolune (BHT) in our crackers and monosodium glutamate (MSG) in our soup.

An Empty Harvest?

Is the food produced by modern agricultural methods good for us? Most plants require between 14 and 20 elements for healthy growth and development, elements which are almost precisely the same elements needed by the human body. Conventional agricultural methods oversupply between three and five elements and neglect the rest.

If these elements are not present in the soil and not taken up by the plants than they cannot be ingested by the people or animals who eat the plants.

To quote the book *Empty Harvest*:

> *A few of the trace elements necessary for plant health and resistance are magnesium, zinc, iron, copper, calcium, boron, manganese, molybdenum, cobalt and chromium. But the absence of one element from the soil can cause great health problems. For instance if inorganic cobalt is missing from the soil the plant cannot absorb it and convert it to organic cobalt. Without organic cobalt the human body cannot manufacture Vitamin B12. When we don't get enough Vitamin B12 we can't assimilate iron properly or make strong red blood cells; we become anemic. Anemic people become weak, depressed and vulnerable to disease. (Jensen and Anderson, 1990, pg. 8)*

Another crucial element needed in microscopic quantities is iodine. The absence of a few molecules of iodine can produce a cretinous child or an adult with grotesquely swollen glands. This particular lesson has been learned well and iodized salt is sold in most inland nations.

But what about the other eighteen substances needed for a healthy life? What are the consequences for us when they are absent from our food?

The modern farmer often regards plants as miniature factories which when supplied with raw materials will produce salable goods. Potassium, phosphorous and nitrogen (PKN) are applied to farmlands generously in the form of chemical salts and a few cheap trace elements are sometimes added if the crop is obviously ailing.

The food produced in this manner looks good. Sometimes it even tastes good. But it is lacking on a molecular level. Some elements are oversupplied, others are absent. Such produce often has elevated levels of sodium and water.

We eat this mineral poor food and feed it to the animals we raise for meat, milk and eggs. In consequence much of what we eat is simply not sufficiently nourishing which causes an avalanche of health problems including obesity.

Overfed and Undernourished: the Multi-Billion Dollar Diet Industry

Obesity has reached the proportions of an epidemic in the United States. Seats in planes, cinemas and buses have been redesigned to accommodate larger people. Every second person is overweight and one in three is obese. As if mocking reality the models and the movie stars get thinner and thinner. How is it possible for these millions of fat people not to hate themselves? They do and try to do something about their weight. But most of them will fail because it is the poor quality of the food in the United States that is causing most of the weight problems.

On the surface this does not seem logical. Obviously obesity is caused by eating foods that are too rich. Doesn't the developed West have the best food in the world?

The food may be rich in calories, fats and sugars but it is generally deficient in minerals, vitamins and enzymes.

The presence of enzymes and trace elements affects the appetite. A person with a healthy appetite becomes hungry, eats and is satisfied. This person will not be hungry again for several hours and his body weight is usually stable.

Many obese people have appetites that cannot be satisfied. They are always hungry and will wolf vast quantities of mineral and vitamin poor food in an vain attempt to satisfy their hunger. Their weight gain comes from the sugars, calories and fats they ingest and from the retention of water in their bodies. Ironically these overfed individuals are actually starving for the missing elements even as their bodies fatten.

Other very fat people eat very little. But their metabolism has been turned so low by dieting, empty food and a life style that allows for little physical exertion that they will remain fat on amounts of food that should barely sustain life.

A huge incredibly profitable industry has been built around what is actually a convoluted form of malnutrition.

The other effects on human health of a diet of poor quality and over refined foods are all too obvious. Increasing numbers of citizens of the developed world suffer from heart disease, high blood pressure, autoimmune diseases and cancer. These afflictions have been traced to diets low in fiber and high in sugar, carbohydrates and the wrong kinds of fat.

Clearly the issue of food quality is as critical as that of food quantity. It must be addressed in any serious discussion about hunger.

Ecological Concerns More people would like to know the true price of what they are eating.

Is contaminated ground water, salinized land, air pollution, deforestation and environmentally induced disease the inevitable result of modern farming? Are some people dying of hunger because others are feeding their meat animals on grain? Is there some way we can still feed the world's population without spoiling the world?

The fact that these questions are being asked by the consumer had led to the establishment of a new kind of product: one that purports to be ecologically friendly. Much to the astonishment of the retail merchants a significant percentage of consumers were willing to pay extra for ecologically friendly products and most supermarkets now sell some of them.

Erosion of Food Diversity Not as public a concern but just as critical is the steady narrowing of the crops used in the manufacture of prepared foods. It is possible to be misled by the apparent variety available while closer examination of the labels reveals that many seemingly different foods are made out of the same ingredients, i.e. wheat flour, corn syrup, citrus, soy with the addition of chemical flavorings and colorings.

There are troubling indications that the over use of these foodstuffs is directly linked to the numerous allergies, sensitivities and adverse reactions that plague the citizens of the United States. Certainly no other nation pays so much to eat such an impoverished artificial, unhealthy diet. Consequently, in the rest of the world, food allergies are still relatively rare.

Genetic Erosion Out in the fields the diversity of centuries of crop cultivation has melted away and modern farmers are all growing the same varieties of plants in immense unbroken stands, a situation so unnatural that only way to keep these plants from being devoured by pests is to use increasingly powerful crop protective chemicals.

The same process is happening in the fruit and vegetable markets as heirloom varieties and regional favorites are crowded out by the McDonald's potato, the Star-Ruby grapefruit and the Delicious apple.

The exception to this trend is the interest and demand for exotic fruits, greens and mushrooms but these occasional treats cannot make up for the steady consumption of fruits and vegetables deficient in vitamins and minerals.

Genetic Engineering/Genetic Modification (GE/GM) This new technology has brought new worries and new questions to agriculture. Are plants carrying the genes of unrelated organisms fit to be food? Will these alien genetic sequences escape into the gene pool via pollen or escape via seed into unregulated ecosystems? How will the other creatures that come in contact with GM plants be affected? Will our health or welfare be endangered by food which has been altered at its most basic level?

The Relentless Spread of Hunger Despite the efforts of international aid organizations and the programs of hundreds of NGO's (non governmental organization) there are more hungry people every year. Technological advances, relief shipments and emergency air lifts seem to have no effect on the swelling ranks of people who cannot make enough money to buy food and do not have land on which to produce if for themselves. There is even evidence that these heroic measures are making the problem worse.

We are presented with many unanswered questions regarding our food supply as we enter the new millennium:

Are there better ways to help the hungry?

Is genetic engineering enhancing or endangering our efforts to feed the world?

Is food produced by modern agricultural methods undermining our health?

Are there better ways to feed the planet than modern farming methods?

Will there be enough food for our children and grandchildren?

How can we achieve true abundance?

Until recently all food was produced by traditional methods. It was all "organic" since farmers added compost and manure to the soil and pesticides and herbicides had not yet been invented.

Farming traditionally, nine percent of the food was destroyed by insects or pests. Food was sometimes scarce and always expensive.

Artificial fertilizers and other chemicals appeared in the last century and seemed to change this picture.

In 1855 the German chemist Baron Justus von Liebig published a book called *Agricultural Chemistry* (von Liebig, 1855, Introduction) in which he declared he had analyzed plant ash and now knew what elements plants needed to grow. This was the origin of the PKN school of agriculture as the elements he decided were necessary were phosphorous (P), nitrogen (N) and potassium (K). His book was seized upon as a sort of Bible by the German chemical industry and "artificial manure" was soon exported all over the world.

On the surface the chemical "manure" looked like a solution to a very important problem. There was simply not enough animal manure and composted material in the 1800s to support the world's growing population.

But Liebug knew nothing about trace elements, fungus and microbes. He did not realize that the balance between mycorrhiza and soil bacteria was a healthy plant's first line of defense nor did he realize that these minute soil symbionts acted as a living bridge between the inorganic world of minerals and the tissue of the plant's roots.

After three decades of experiments with food plants he did realize that he had launched a very destructive science on the world and later tried to correct his initial oversimplified explanations of plant metabolism. It was already too late. The first bubbles of artificial abundance were appearing, big beautiful bountiful crops of fruits, vegetables and grain which actually had little food value.

Opposition to chemical based agriculture was suppressed or relegated to faddist circles. This suppression was so complete it is still against the law in much of the West to claim that organic produce is superior in food value. In the United States it is not legal to claim that organic produce is "healthier."

Considering what claims are made on the media by the makers of everything from cars to cigarettes this is a very strange situation.

As the yields increased, quality declined and the soil, subjected to repeated applications of chemical salts began to die.

The balance between soil bacteria and mycorrhiza could no longer be maintained and new plant diseases sprang up in the artificially lush field while old ones gained a new virulence.

The next step was the development of crop protective chemicals, all of them poisons, to keep the weak, water filled, mineral deficient identical plants from being destroyed by insects, fungi, bacteria, viruses and blights. These chemicals came into use worldwide in the latter part of the 20th century.

As a new century begins crop loss from insect damage has doubled since WWII to a staggering 18% while the use of crop protective chemicals has increased by 3000%.

The "pesticide spiral," proposed as a possibility by Rachel Carson in her book *The Silent Spring* (Carson, 1962, pg. 217-231) has become a reality as more chemicals are used with diminishing effect on increasingly resistant pest populations. The chemicals are causing severe health problems and great ecological damage. Using anything more lethal will make the food too poisonous to eat.

GM proponents claim that plants can be engineered to secrete their own pesticides.

This school of thought ignores the fact that insect populations are endlessly adaptable; *they will soon be resistant* and we are not.

Pesticide producing plants secrete toxin from every cell through the plant's entire life cycle. This toxin cannot be washed off and is ingested when the plant is used for food. The toxin kills beneficial insects as well as problematic ones.

To find our way out of the chemical dead end we have to think back to Liebig's initial mistake. Healthy food can only come from healthy soil. Healthy soil is living soil full of worms, bacteria, fungus and the full panoply of minerals.

Agriculture Must be Corrected from the Ground Up.

The Sides of the Pyramid *From the Ground Up: Rethinking Industrial Agriculture* (Goering et al, 1993) is actually the title of a book that has served as a primer for the proponents of sustainable agriculture.

In one of the opening paragraphs the authors state: *This reductionist perspective has allowed the earth under our feet to be seen as little more than a factor of production, soil as more or less inert matter, ecosystems as resources to be exploited. In mainstream circles the notion of a living earth receives lip service at best: the thrust of modern society is the de-animation of the world.* (Goering et al, 1993, pg. 3)

This attitude prevails despite the fact that the life of the soil can be weighed, measured, and observed. A teaspoon of living soil contains millions of tiny organisms and the presence of these organisms is necessary if we are to raise healthy food crops.

The authors continue: *This narrow scientific view goes hand and hand with an equally narrow economic paradigm. Modern economics is concerned almost exclusively with factors which can be reduced to numbers, GNP, the principle yardstick now used to measure national wealth, is based solely on the exchange of money. Non-monetary transactions—whether in the household in industrialized economies or with subsistence economies in the South—are ignored. So too are all the non-quantifiable factors such as sustainability, social cohesion, psychological well being and aesthetics. The narrow focus* *on economic growth allows a whole range of problems to be accepted as the "price of progress." The breakdown of family life, individual alienation, spiritual poverty and the destruction of the community are all seen as mere side effects of the unquestioned push to increase the GNP." (*Goering et al, 1993, pg. 4)

The narrow view also allows modern agriculture to masquerade as one of the great achievements of the industrial age while many people are starving, the soil is dead or dying and farmers are being driven off the land by the millions.

This is because the modern scientist and economist has misunderstood the nature of agriculture, making as profound an error as Liebig did when he declared that only three elements out of the necessary twenty could support healthy plant life.

Agriculture is a cooperative effort, an orchestrated symbiotic arrangement between the living soil of the planet, the action of natural forces, the plants and animals being raised and the people who care for them.

If one looks closely at modern agriculture with its eroding and poisoned soil, polluted water, antibiotic-saturated animals, poison-sprayed crops and bankrupt farmers one realizes that all the living, interacting, productive elements are in dire straits.

The Gross National Product (GNP) may be increasing: but the GNP does not measure the well-being of anyone or anything.

The ability of the human population to raise up food out of the earth is in desperate decline.

Increasing the Food Supply

All of the living factors of the agricultural pyramid must be strengthened to reverse this decline. These measures include restoring soil, preserving the quality of water resources, stopping genetic erosion of crop species and most of all protecting the rights of the farmers.

Bringing new crops into cultivation or restoring lost crops to the agricultural mainstream may be the quickest way to increase the available supply of food for farm populations.

There are several reasons why this is so. Many of these crops are not considered important enough to "own." No corporate giant has "improved" them. No research organization has claimed them. There is no fee for their use. The farmer needs only a handful of seed and his or her own skill.

These crops have not been subjected to the debilitating breeding programs that have robbed conventional crops of much of their vigor. They are still genetically diverse and so genetically flexible. They retain much of their natural resistance to insects and plant diseases.

Many of the crops can be planted in marginal areas, the kind of sparse, harsh or dry areas that cannot sustain conventional crop plants. These hardy plants are ideal for the reclamation of areas depleted by conventional agriculture.

On the economic level there arguments against the kind of grass-roots renewal I am advocating such as the lack of funds for organized research and the limited markets for unfamiliar foodstuffs. There are also no organized systems that handle these crops for they are not considered to be valuable.

I do not propose that these crops go immediately into commercial production. I suggest that they be grown for private and local consumption first, entering the economic system via the household and the farmer's market.

It is only on the private and local level that we can see the true value of these plants as they relate to the well-being of the living elements of the agricultural pyramid.

Asking the Right Questions

Good food is a **priceless** commodity.

The so-called "market economy" only involves the developed world. Seventy per cent of the food produced in the world will never get to a shop but will be eaten a few meters or a few miles from where it is grown.

"How much can you get per pound for these things?" is the wrong question reflecting the bottom-line mentality of modern commercial agricultural

The questions we ask should relate to the health of the people, the plants, the farm animals and the soil. If the living elements of the agricultural pyramid derive sustenance and benefit from the plant's cultivation **the market value is irrelevant.**

If the presence or cultivation of the plant supplies the needs of farming populations for food, fodder, energy, building material or medicine, just about **every other kind of investigation, recognition or sanction is irrelevant.**

The needs of farming populations are inevitably neglected by the market economy. Plants that are good for these populations are precisely the plants that are needed.

Should we Really be Taking These Plants from one Place to Another?

People have been migrating and bringing their food plants with them for thousands of years, sometimes with unfortunate results. The vast majority of the new crops and crop candidates in this book are not suspected of having invasive or weedy traits. (See Appendix 4 on page 339 for the definitions of weediness.) Those plants, which might be invasive, are labeled as such and should be handled with care.

As a general rule plant introduction is less risky than the solutions put forward by conventional agriculture for increasing the food supply, most of which advocate genetic engineering, increased chemical use and other expensive high tech blind alleys.

What Should We Try First?

Grains Grains are the source of eighty per cent of the food energy in the world. Even one new grain crop in a marginal area can make a difference in the quality of life. The widespread cultivation of new grains could bring the world closer to food security

Oil Seeds Oils are vital as an energy source, and are also used for cosmetics and for cooking. Commercially produced oils are often adulterated, tainted or of poor quality in the developing world. Edible oil production generates local employment and oils have high trade value.

Trees Crops of all Kinds Arboreal agriculture is generally low input, low energy, anti-erosive, soil regenerative and high yield with the added rural benefit of seasonal employment. Value added products such as dried fruits, preserves, jams, and teas can be produced with little investment. Orchards also provide cover for vegetable and fodder crops as well as firewood.

Greens Vitamin rich foods squeezed out of commercial agricultural systems for "economic" reasons, grown locally and consumed locally are a much cheaper and more sensible solution to the problem of Vitamin A deficiency in Asia than the "golden rice" initiative. Many of these exotic greens, which the rice farmers cannot "afford" to grow any more, have recently shown up on Western tables.

Nuts Nuts have excellent food values, they require low input, low energy cultivation. Nut trees are generally long-lived. Nuts are storable with a high trade value.

Spices and Tertiary Foods Spices are small cash crops for farmers and farm communities.

They are easily transported, have high trade value and high food value. Their production and packaging generates employment. An area's cuisine is defined by the way spices are used.

Exotic Fruit Exotic fruits are high vitamin food with high trade value and the possibilities of value added product production.

Another question we should ask:

How can these plants be cultivated for the mutual benefit of soil, farmers, farm animals and the plants themselves?

The need for diversification

Tunnel Vision Investors with a weak or monolithic stock portfolio are often told to diversify. In every day life we are sometimes advised "not to put all our eggs in one basket."

We all "have stock" in the world's food supply and many of us can see that the overproduction of twenty genetically impoverished mega crops has made our portfolio feeble indeed.

The floods of cheap staples then depress the local staple markets forcing out all but the biggest producers. The giants then compete with each other and the price drops further while the genetic base of the crop narrows and the use of chemicals and machinery skyrockets. The "big" farmer in the developing world is caught on a treadmill on which more and more of the same product has to be squeezed out of the land for less and less return. In developing countries the agricultural populations are forced into growing cash crops, the raw materials the developed nations will use to generate wealth, while their own food needs are neglected.

Since all the living elements of the farm are suffering under such a system it requires tunnel vision to call this "good business."

In fact, a controversial book claims that the United States of America (US) and the modern system of agriculture is "exporting famine."

The Hunger Machine In *The Hunger Machine,* co-author Susan George describes the darker side of development. (Bennet and George 1987, pg. 2)

> *Most of the policies which the rich world has practiced vis a vis the South has been destructive for their peasantries and worsened the hunger problem. Techniques that work in Iowa are not a prescription for radically different societies. Strategies like the "Green Revolution" have contributed to concentrating land in fewer hands leaving millions with no place to go. Transnational agribusiness corporations may increase their own sales but they also increase the price of food in countries where the majority is far too poor to buy it. Food aid may have saved some lives for a time but it has also made countries more dependent on imports, on further handouts. Huge "development" projects have tended to benefit the elites, increasing their capacity to oppress the poor and the hungry. Sometimes in the case of large dams, such development simply displaces thousands of peasants without compensation, and they without alternatives, except migration to mushrooming cities where they find no jobs.*

More injustices and inequalities exist in hungry countries nested one inside the other like Russian dolls. Urban populations get more food and food aid than rural populations. The World Bank estimates that 90% of the hungry live in the countryside. In other words people who produce food or could be producing food are the first to feel the lack of it. (Bennet and George, 1987, pg. 6)

Within the rural population there are also inequalities between those who own the land and those who work it. Laborers have access to approximately one third of the amount of food of the landowning peasant because of the seasonality of rural employment. Within the rural laboring class men usually get 29 percent more food than their wives and boys 16 percent more food than their sisters.

Paradoxically the female rural laborer is the worst off of all. Women work two thirds of all the hours worked in the developing world, they get one tenth of the income and own one hundredth of the property. It goes without saying that the rural population also gets less education and medical care than the urban population. So it is no surprise that 90 percent of the people who die of hunger are from the rural poor. (Bennet and George,1987, pg. 9)

> *The circumstances in which these people live are as varied as the many cultures of the human race; yet the hungry have one thing in common—they are all poor. It is poverty that starves people to death, not the callous whims of nature, nor even the stupidity of war. And the overall war against poverty is being lost: there are more hungry people in the world today then there were a decade ago. (Bennet*

and George, 1987, pg. 15)

A farmer in the US, a farmer in the Ukraine and a farmer in Bangladesh may seem to be worlds apart but are they held in contempt by "business," considered expendable by their governments and slated for extinction by economists. Wave after wave of foreclosures and bankruptcies have depopulated the breadbaskets of the West. The situation is so bad in the former Soviet Union that the Ukraine, an area that once supplied Europe with high quality wheat, cannot feed itself. In Bangladesh and other very poor nations it is not merely financial ruin and hunger that awaits the farmer.

Slavery has made a big comeback in the twentieth century. According to the editorial of *International Herald Tribune* on Sept. 11, 2000 (reprinted from *The New York Times*) there are 27 million slaves in the world, two thirds of them impoverished rural laborers held in "debt bondage," a system which entraps whole families into peonage to pay debts incurred by medical expenses, crop failures or funerals.

> *Slavery and related kinds of servitude are a growing business because the number of desperately poor people is increasing and globalization has disrupted rural communities. (International Herald Tribune, 2000)*

It is worthwhile to note that the fiercest opponents of "globalization" are the poor and populous nations, which already find they are powerless within a system of injustice. They fear that their local and national priorities will become even less important when "good business" which has devastated the farming systems of the world extends to other realms of endeavor.

Localization Instead In the modern world the less one has to do with real work, the more one is paid. In the modern world the closer one is the land the less likely it is that one will enjoy the wealth generated on it.

In the modern world the more "globalization" improves the world's economy the more we see of withering communities, collapsing societies and man-made desolation. But globalization is not a sinister conspiracy but only the last step in the wrong direction. The rules and policies that have leached away the vigor of the countryside have been in practice for over a hundred years.

Logically the only antidote to this madness is to concentrate on improving local economies. Instead of a "trickle-down" theory or a Global Marshall Plan we need a plan that will create abundance where the food of the world is actually grown.

Diversification at the local level on poor land, depleted land and marginal land will create more food for the farmer, the farm family and the farm animals. The revival of old crops will preserve in-crop diversity. Concentrating on local needs will mean that the farmer may again acquire locally produced building material, fertilizer, clothing and value added products. New crops can also be the answer to the "hungry seasons" before the harvest. Much of this bounty will not register on the GNP, it will be given in trade or passed about among family members, but the farmers will use it and benefit from its use.

Appropriate technology will bring light, heat and communication to isolated communities breaking the rural neglect and paralysis inherent in urban based centralized systems. They may be able to supplement their incomes by "farming" energy from sun or wind and selling it to the local grids. Rural people will be on the internet, setting up fair trade networks and community supported agricultural projects, involved in cooperative ventures, planning for the future, looking out for the interests of the community.

The living, interacting elements of the farm that suffer so under the modern system are the plants and animals, land and people. Ensuring their well-being and kindly "use" would do a great deal toward creating not an artificial bubble of false plenty but a healthy surplus of nutrition.

Abundance will radiate outward from productive farms to the small communities that serve them. There will again be jobs for the young people of rural communities, stopping the flight from the countryside. There will be good food in the marketplaces. The food will be consumed and bought, sold locally and the local market will determine the nature of the produce and its price. Such food will be inexpensive because it is not wrapped six times in plastic, processed into expensive inferior products or flown in by jet plane. A possible source of jobs income and employment would be taking this healthful local food into the nearby cities and selling it in areas which food stores have abandoned, benefiting both the rural and urban poor at the same time.

Seventy to eighty per cent of the food available in a community should be locally produced. This ensures that it will be fresh, relatively cheap and there will be consumer control over the quality. It will also eliminate energy wasted in transport

Twenty percent of the food can then be imported, paid for by local surplus that is made into value added products or sold in other locales. Obviously some of this imported food will be out of season, some will be luxury foods and some will be tertiary foods like spices and tea. These foods should be produced in other healthy local economies bypassing the cartels, monopolies, middlemen and overheated commodities markets that suck all the wealth upward and consume it like a furnace.

If local economies are healthy a nation has a sound citizenry and a solid tax base to promote education and insure medical care and pursue national priorities.

The global economy then takes care of itself.

This, of course, is the exact opposite of the current market system in which the farmer is forced to squeeze the land dry, the wealth of the earth and the human capital are drained out of rural areas and the poor nations' natural resources are bled away to the wealthy countries.

A dismaying illustration of what this drain means on a local level: In North America in 1996 farmers received four cents out of every dollar of food wealth they produced—twenty percent went to the government, 21 percent went to the retailer, 26 percent to the food processors and 29 percent to brokers traders and shippers. (Gene Wars, *The Politics of Biotechnology,* Dawkins, Seven Stories Press, 1997, pg. 15) In other words in 1996 a farmer had to produce 104 pounds of corn to be able to buy a 25 ounce box of cornflakes or 93 pounds of potatoes to be able to order a dish of potatoes with cheese at a local eatery.

Several rounds of agricultural agreements, negotiations and treaties since then have actually made this situation *worse.*

While the systemized decline and impoverishment of rural America is a tragedy the global ramifications are a looming disaster that threatens the food security of the entire world. Prices for the consumer continue to rise even as the share of the actual producer of food continues to diminish.

The horror of systemized impoverishment embodies itself at a global level in "strategic adjustments" and the World Bank advice to countries with outstanding debt. The World Bank panacea for debt is simple: **devalue the currency, sell off assets, reduce the amount of money spent per capita on health, education and welfare.** In other words hurt the citizens by making their savings worthless, taking their jobs and stone-walling their attempts to do anything to better their lot. These measures lock impoverished populations in poverty, ignorance and ill health and can destroy the entire middle class of a nation.

Now entire nations can be "mined" for short term gain rather than mere farming communities increasing the ranks of the poor and hungry by millions.

These same nations are told that they should encourage their citizens to have fewer children.

But birth control is not a solution to the problems of hunger.

In the AIDS-devastated rural areas of Africa the population is plummeting. This does not mean that there is more food to go around. More food can only be produced by more farmers who are working on the land and caring for it. This is precisely the reverse of what is actually happening.

There will be no change for the better until the fundamental errors in how we value, use, sustain and distribute food wealth are addressed.

According to an article released by the Associated Press (AP) on Sept. 13, 2000 the United Nations (UN) Millennial Summit was attended by the representatives of 189 countries who will try to find solutions that will "lift billions of people out of poverty." (Associated press, 2000) Working groups formed at the summit will continue their discussions and activities until 2005.

This is a wonderful goal. Unfortunately many summits attended by hundreds of thousands of educated people in expensive suits have been held before with similar aims. The most likely result with be another unreadable mountain of paper.

Will any summit address the government-sanctioned devastation of the breadbaskets of North America?

Will any summit acknowledge the fact that modern agricultural practices and modern economics are part of the problem, not part of the solution?

Will any summit somehow acquire for rural communities their fair share of the national budgets for health and education?

Will poor farmers in the developing world be able to receive credit on fair terms?

Will subsidies for water and fuel be eliminated so that the true price of conventional agriculture becomes visible?

Will laws be passed so the seed giants and gene giants of the world will be held liable for the environmental damage they cause?

Will laws be passed so that indigenous people may keep their lands, their trees, their right to support themselves?

If the answers are no—future summits should be canceled, the money slated for use in organizing, traveling, translating and speechifying should taken and given to the mothers of hungry children in the poorest nations on earth.

They will know what to do with it.

What We Can Do

> For we who love the land and live and work on the land no government program or international summit can help us. Even the well intentioned among the experts do not know how to come to our aid. *We must help ourselves, save ourselves, aid ourselves, educate ourselves.* Whether we stay in business by abandoning a futile attempt to compete with the gene giants and start to grow tef instead of corn or open a bed and breakfast to bring eco-tourists to the farm—the solutions that we find must be sustainable, healthy and affirming of the bond between us, the land, the crops and the animals. Imitating the same systems that are making the earth uninhabitable will get us nowhere.

That brings us back to the base of the pyramid. We need new crops, new methods, new ideas, new forms of cooperation to keep us fed, employed, solvent *and on the land.*

Increasing the Food Supply by Diversification

There are at least 15,000 edible plants in the world. Only 150 plants are widely used as food for human consumption. Fewer than one hundred plants are cultivated in commercial food production. In developed countries the number of plants species being cultivated has declined even further and in the name of efficiency and mass production immense tracts of land are planted with identical stands of the same few staple crops.

Author Collin Tudge comments on the narrow genetic base of modern food production in his book, *Food Crops for the Future.*

> *There are about 250,000 species of flowering plants and according to the National Academy of Sciences human beings in their time have exploited about 3,000 of those species for food and cultivated about 150 on a commercial scale. However by far the largest amount of food for human beings is now being supplied by only around 20 different kinds of crop. (Tudge, 1988 pg. 10)*

Why Diversify?

We are told that the global population is going to increase by 2050 to nine billion. Where will the food to feed the extra billions come from? If we can manage to produce the needed food will there be room for anything on the planet but humans and our cultivated plants?

Mainstream agricultural science offers us genetic engineering and factory farming as a solution to the problem. Supposedly corn, wheat, soybeans and other food plants will be "improved" with the additions of genes from dozens of unrelated organisms. Farming will become even more "efficient" and dependent on machinery. Immense yields of inexpensive GM food will come from fields that are surveyed by satellite, tilled, planted and harvested by remote control.

This sounds dismayingly like the promises of the proponents of nuclear power. Fifty years ago many voices were raised against the "peaceful" uses of the atom arguing that splitting the atom to heat water to turn steam turbines to produce electricity made about as much sense as burning down your house to cook a pan of sausages. The proponents of nuclear power prevailed. They promised unlimited energy for all, so cheaply that it would be given away.

What they delivered was a waste problem that will haunt humanity for thousands of years, an increase in deaths from radiation induced disease and numerous sites like Chernobyl and Three Mile Island which have to be sealed off from the rest of the world forever.

I do not want to trust the future of the human race to the mindset that brought us DDT, Round Up, Bt potatoes and Terminator Genes. Nor do I see much of a future for factory farming which uses approximately 28 calories of nonrenewable energy to produce four calories of dubious quality food.

To feed our children and grandchildren we need a new kind of agriculture. What type of agriculture this will be is still not clear but more pieces of the puzzle are falling into place day by day.

Farming for the New Millennium The new agriculture must be diverse both in cultivars and within cultivars to prevent the terrible buildups of parasitic organisms that plague monocropping.

The new agriculture must neither destroy the soil, air and water nor genetically impoverish the organisms on which we depend for our living.

The new agriculture must not use more energy than it produces. Modern agriculturalists often sneer at the subsistence farmer but at least the latter does little harm.

The new agriculture must leave room for other living creatures on the face of the earth. *Efficient cultivation—by which I mean, specifically, cultivation that produces* good *food but does not obliterate all living things—is a sine qua non if we are to survive in a tolerable world."* (Tudge, 1988, pg. 1)

I would like to take this line of reasoning a little further. Truly efficient cultivation is a *sine qua non* if we are going to survive, period. To survive as a population of nine billion we will need clean water, rich soil, breathable air and very large quantities of high quality fruit, vegetable and grain based foods, as well as lean chemical free meats, fish without heavy metals contamination, safe dairy products and pure edible oils.

To achieve this we should be investigating every edible plant on the face of the earth.

Why Haven't these Plants Been Used as Food Before?
Some of the reasons these plants have not been used as food before are purely techni-

cal. Large seeds are easier to gather than small seeds. Seeds that separate readily are easier to mill and bake than seeds which have to be decorticated. This is why wheat, corn and rice are preferred grains and millet, tef and sorghum have remained subsistence crops.

Food preferences among people account for some of the difficulty in introducing new crops. Raised breads are desirable food, especially for the urban poor who often get much of their food energy from bread. Consequently many poor countries spend a large percentage of their budget on imported wheat while their indigenous cereal crops are underutilized or ignored.

Some plants once cultivated as food were "lost" to cultivation as the people who valued them went into decline or the conditions that encouraged cultivation changed.

Political and economic reasons cannot be ignored. Dominant cultures take crops they desire from weaker indigenous societies, force their preferred cultivars on the natives then relegate the unwanted plants to the status of "famine food" or "poor people's crops." Until recently this attitude insured no serious research would be done on a cultivar hence it would never have any market value.

As we begin the next millennium of agricultural endeavor all this must change.

To achieve true abundance and a tolerable world we must embark on a true exploration of the food resources available. We may soon need every relic crop, heirloom variety, native cultivar and ancient grain to nourish ourselves, our children and our descendants.

This is by no means an encyclopedia of little known food crops but only a taste of what is "out there" if we are wise enough to find, cherish, cultivate and use it.

Lost Crops and Underutilized Plant Resources

Little known plants which can regenerate depleted land, nourish animals and people, supply fodder, building material and many other useful substances are described in the following books:

The Lost Crops of The Incas, National Academy of Sciences

Underexploited Tropical Plants With Promising Economic Value, National Academy of Sciences

Food Crops For the Future, Colin Tudge

Enduring Seeds, Gary Nabhan

The Lost Crops of Africa, vol. I, Grains

The Lost Crops of Africa, vol. II, Fruits and Vegetables, National Academy of Sciences

IV. Lost Crops

Reviving lost crops might be the easiest way of providing more food and necessary materials on farms (which in many places is exactly where they are most needed) because so much of the hard work of domestication has already been done. In many cases the lost crops are not really gone. They have been shoved aside into local backyards because a commercial variety of the plant has taken over the farms. Or the plants exist in the working collections of seed savers or survive in the plots of subsistence farmers. Now is the time to save them, revive them and put them back to work on the farms before they become completely "lost" to us.

What is a "Lost" Crop?

A lost crop is a species or even a race of plants that has already been domesticated by farmers and agriculturalists but for one reason or another is not used or underutilized today.

Who Lost it?

Sometimes lost crops were merely displaced by plants that were more valuable economically or better adapted and the process was gradual and gentle as new cultivars replaced the old. This is what happened with the most primitive diploid wheats, which were replaced over a period of centuries by tetraploid and hexaploid strains, which were easier to thresh, mill and bake.

Sometimes the crop was deliberately eradicated as a way of attacking the people who used it.

The Roman legions that destroyed Jerusalem were ordered to take a long, dangerous and costly detour southwards to the shores of the Dead Sea. There they were deployed around a grove 2,000 sacred and venerable basalm trees, the only such grove in the world. The trees were the source of oil used in the Temple. Taking these trees was the Roman way of striking at the spiritual side of the rebellious Jews. Evidence at Ein Gedi in modern day Israel tells an amazing story. The Romans did not know how to manufacture the sacred oil or the perfume. The ancient Jewish community was spared the fate of the rest of "Palestinian" Jews and was allowed to go on cultivating the trees and making the precious perfume. The synagogue at Ein Gedi was in use until the sixth century A.D. when the trees and the Jews of Ein Gedi both disappeared from the landscape.

Hernando Cortez was so determined to destroy the grain amaranths that were the staplefoods of the Aztec empire that he ordered his soldiers to behead anyone planting the grain and cut off the hands of anyone caught using it. Amaranth was central to many festivals and rituals of the Native Americans. It was precisely amaranth's "extra" dimensions that were so hated and feared by the Spaniards. (Cole, 1979 pg. 18) While the Spaniards did not manage to eliminate amaranth the ban on cultivation did keep deprive the world of a major food crop for several hundred years.

Changes in economic priorities cause crops to become "lost."

Often "lost crops" were simply forgotten as cash cropping replaced the more diverse and productive indigenous systems that had fed the people of the developing world for thousands of years. Good examples include many "heirloom" varieties of common fruits and vegetables, special plants with special properties squeezed out by a foolish and unnecessary drive towards standardization

Finding Lost Crops and Bringing them Back into Production

Whatever the reason for the original eclipse of these botanical treasures there is no excuse for their neglect now.

The following plants are from locales as diverse as the Moroccan desert and the jungles of Central America. They include legumes, cacti and trees. Some are well along the path to domestication. Others are not far genetically from the wild stock they were taken from. What they have in common is that they have all been grown at some time in an organized manner in their native ranges. Then for some reason they never entered the mainstream of botanical and agricultural science.

Needless to say these old/new crops can be grown in marginal areas, contain fresh genetic material and have unique characteristics as food plants.

Basul: The Tree-Bean of the Andes

Basul (*Erythrina edulis*) trees are small evergreen trees with lush broad-leafed foliage that have been cultivated in the Andes since pre-Incan times. They are vigorous and fast growing, often the first tree to colonize newly cleared sites. In its native range basul is esteemed both for its beauty as a boundary tree and living fence and its usefulness as a source of food. Basul is a legume and a true utility tree as it produces edible seeds and flowers for human consumption, nitrogen-rich leaves which can be used in mulch or fed to animals, firewood, lumber and poles for construction.

As a leguminous tree it enriches the soil around it. Basul trees can be planted as windbreakers and used to shade more delicate crops.

Basul pods ripen before field crops are ready providing an important bridge crop for poorer areas where seasonal food deficits are common. These beans can eaten the same day they are picked after being boiled in water or dried and saved for future use. They should not be eaten raw. The beans are tender and tasty like large lima beans and nutritious. Basul beans contain twenty percent protein by dry weight and generous amount of lysine and phosphorus.

Local Importance: Famine-breaker, bridge crop and inter-crop. Reliable source of protein. Source of wood for construction and fuel. Fodder for food animals.

Potential Importance: New food source for zones of Africa and Asia. Land reclamation and "nurse" tree for delicate plantings. May be suitable for agro forestry in warm areas with sufficient water as basul is not frost or drought tolerant.

Comment: Particularly suitable for subsistence agricultural, land reclamation and agro forestry because of basul's beneficial effect on other plants.

Tarwi: The Soybean of the Future?

Tarwi is a species of lupin, (*Lupinus mutabilis*) sometimes called Andean lupine. It is an erect annual with a hollow stem and a deep taproot. Its beautiful blue flowers make it one of the most striking of cultivated crops. It has been grown for food in the uplands of Peru, Bolivia and Ecuador for thousands of years but only recently attracted the notice of botanists and agronomists.

Tarwi's seeds are outstandingly nutritious with 40 percent protein, as much as soybeans and peanuts. Tarwi seeds also contain up to 20 percent oil, as much as most commercial oilseed crops. The bone-white seeds also contain lysine and cystine (deficient in most cereals) and generous amounts of linoelic acid.

These qualities make tarwi a potential source of human food, animal feed and edible oil. Research may reveal many other uses.

Tarwi seeds are served cooked in soups stews or salads much like chickpeas. They can also be made into a poi-like paste.

Leaching and cooking the seeds is necessary before human consumption as the seeds contain bitter alkaloids.

Tarwi is also a pioneer plant that can be cultivated on marginal soils and used to enrich depleted land. As it is a legume it fixes nitrogen and its deep roots loosen compacted or rocky soil.

The plants are very adaptable to a wide range of soil, are drought and cold tolerant and unlike many Andean plants do not require any specific photoperiod.

Tarwi's growing season is long and the plants produce many tiers of seeds, which may make mechanical harvesting complicated, but it is a boon for subsistence farmers.

Local Importance: Important source of protein and vegetable oil. Green fodder, green manure, animal feed. Intercrop with corn, beans, potatoes and quinoa.

Potential Importance: Maybe be as versatile as the soybean. The nutritional equivalent of the soybean and the peanut. Land enrichment agent and anti erosive. Possible source of superior edible oil.

Comments: Suitable for integrated projects, subsistence agricultural and sloping lands. It should be evaluated as an oil crop.

Lucuma: Maple Syrup of the Andes

The lucuma (*Pouteria lucuma*) grows in the temperate uplands of Columbia Ecuador Chile and Peru (colorplate 1 on page 59). It somewhat resembles another member of the Pouteria family, the sapodilla or chico sapote, a better-known plant esteemed as an ornamental and a fruit tree. Domesticated by the Incas, its smooth round bronze colored fruit looks like that of the persimmon tree. Its orange and yellow flesh is soft when the fruit is ripe and usually blended into other foods.

It can made into shakes, puddings, drinks, and baked good, jams and syrups. The taste, color and smell is very much like that of real maple syrup though the texture is firm and pumpkin like even in very ripe fruits. Lucuma fruit pulp can be frozen and stored or dried and ground into flour.

Lucuma fruits are high in solids and a good source of calories and carbohydrates. They are also a good source of iron, trace elements and Vitamins A and B.

Lucuma fruits can weigh as much as a kilo and the trees fruit generously all year. This is an important trait in areas where food is scarce before the harvest of field crops. The fruit pulp is also suitable for poultry feed and the tree produces a dense hardwood timber. The lucuma tree does not need any specific day length.

There is some evidence that the lucuma tree is well adapted to sandy and rocky sites. It is moderately drought and salt tolerant but can be killed by temperatures below -5°C and is sensitive to waterlogged soils.

Local Importance: Reliable source of food for poorer areas, nutritious ingredient of processed food, local sweet.

Potential Importance: New fruit crop for frost-free upland areas of the sub tropics and tropics. Could be very useful in Africa and much of South America. A possible cash crop for these areas.

Comments: The size and quality of the fruit is superior to that of many cultivated orchard trees. Better cultivation methods, propagation and breeding are needed to bring lucuma into the agricultural mainstream.

Sapodilla: Don't Judge the Fruit by its Peel

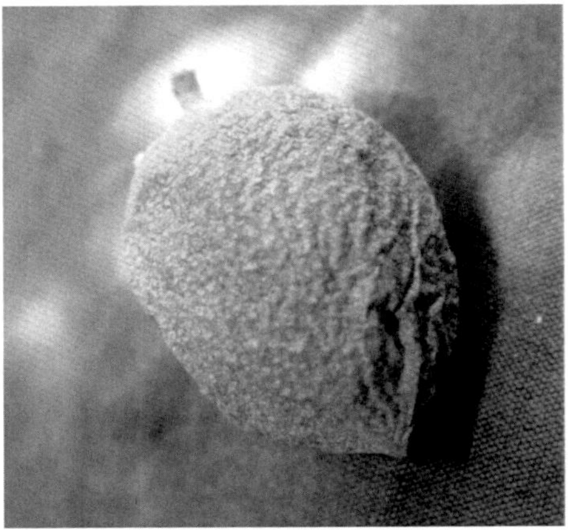

Photo by Elaine Solowey

Figure 1: Sapodilla

Sapodilla or chico sapote (*Manilkara zapota, Achras sapota*) is the fruit of the chicle tree (colorplate 2 on page 59). This hardy broad-leafed evergreen tree produces rough-skinned ovate fruit with brown flesh and the taste of a candied pear. Two to five flat oval seeds cluster in the fruit's core but the rest is deliciously edible. The fruits are as nutritious as the lucuma but smaller and moister. The trees are less sensitive to salt, low temperatures and day length than the lucuma. Sapodillas are grown in Central and South

America and South East Asia as minor fruit crop but could be useful in many areas. The fruiting season for the sapodilla trees is as long as six months in some places, providing a steady supply of salable fruit and nutrition.

Local importance: Fresh fruit, chicle, some timber

Potential Importance: New crop for subtropical areas and the semi arid tropics

Comments: The sapodilla may be suitable for the production of VAP, such as kamaradin, flavoring and preserves

---

The Ugni: A Wild Berry or a Cultivated One?

Ugni berries (*Myrtus ugni*) grow on slow-growing evergreen shrubs. The berries are very much like the myrtle berry in appearance they are sometimes marketed under the name myrtle berry but are larger, sweet and edible with a wild strawberry like flavor. Ugni berries are cultivated commercially in mountains Chile that may be their original range. But the Incas spread the ugni berry into other mountain forest locales as a cultivated plant. There the trees returned to the wild state.

Like the cranberry the ugni is valued for its sprightly flavor and makes pleasantly sharp jellies, jams, drinks and desserts. Ugni berries contain a healthy range of vitamins and sugars and impart a deep purple color to other foods, which could replace artificial food colorings.

Ugnis are also valued for their beautiful pink and purple flowers and their glossy foliage and have become popular ornamental shrubs for city gardens in California and New Zealand. They are easily propagated by seeds, drought tolerant and tolerate some frost. They do not need any specific photoperiod In Chile, some superior strains are propagated by cuttings.

Local Importance: A local cash crop and sweet.

Potential Importance: Healthy alternative to food coloring and artificial flavoring. Possible cash crop for hilly areas. Pleasant new fruit for areas that are not too cold.

Comments: The plant is suitable for integrated projects, gardens and terraces because of its sturdiness, modest size and high yield.

The Cereus Cactus Apple: Nature's Ice Cream

Photo by Ruthi Kubenstein

Figure 2: Cereus Plant

The Cereus fruit (colorplate 3 on page 59, colorplate 4 on page 59, and colorplate 6 on page 60) grows on a handsome columnar cactus from Peru (*Cereus peruvianes*). These fruits were cultivated and prized by the Incas but survived only as backyard food plants, wild plants and ornamentals. The reason for this is unknown since the fruit of the cereus cactus is spineless, beautifully colored and refreshing.

Cerues cacti are usually propagated by cuttings. A large lobe is separated from the mother plant, allowed to dry for a few days in the shade then planted a third of its length in sandy, well drained soil. After two years the Cereus cactus begins to produce large flowers, which are pollinated by bats, bees and flies. Four to six weeks after pollination the Cereus fruit ripens. Its smooth glowing spineless skin may be purple, pink, red or yellow. Some fruits are fragrant. The pulp is white and smooth with small black edible seeds and a cool sweet taste.

Like most cactus fruit the Cereus fruit has few calories and is a good source of Vitamin C.

The Cereus cacti are both drought and salt tolerant but are sensitive to water logging and frost.

Cereus cacti have been planted in Florida and Hawaii as ornamental plants and introduced experimentally to Israel and new Zealand as a crop candidate. In Israel this fruit is marketed under the name "Kubo fruit."

Local Importance: Backyard tree, ornamental.

Potential Importance: Possible fruit crop for arid and saline lands.

Comment: A delightful addition to the garden, useful in subsistence projects and suitable for commercial planting.

Table 1: Cereus Peruvianes –Food value per 100 grams of edible pulp

Calories	42-54
Moisture	83.8-86.1 g
Protein	0.09-0.12 g
Fat	0.8 g
Carbohydrates	10—13 g
Soluble sugars	5.8—8.2 g
Acids (citric)	0.2 –0.5 g
Fiber	1.0—1.5 g
Ash	0.30—0.43 g
Magnesium	13.5—15.8 mg
Calcium	5.2—11.4 mg
Phosphorous	16.9—18.5 mg
Iron	0.28—0.70 mg
Sodium	2.9—4.1 mg
Potassium	63.2—93.4 mg
Vitamin C	31.4—54.0

analysis by Ben-Gurion University

Pitaya or Strawberry Pear: The Spectacular Fruit of the Night-Blooming Cactus

These plants are believed to be native to southern Mexico, Guatemala and Costa Rica. They are a family of viney, sprawling epiphitic cacti that prefer partial shade (colorplate 8 on page 60 and colorplate 7 on page 60). Pitayas (*Hylocereus sp.*) were domesticated thousands of years ago by indigenous peoples of the area and allowed to grow up other valuable plants, using them as a natural trellis.

Many species became naturalized in the West Indies and other islands and were taken to tropical locales in the Old World as well.

Oddly enough it was not their large and beautiful fruits that attracted attention. Hylocereus cacti were esteemed for their huge, fragrant bell shaped flowers, trumpet sized blooms that open at dusk and last only for one glorious night before withering.

The fruit escaped notice because the Hylocereus cacti usually did not bear fruit in its new homes. Many varieties are self-sterile and lacked a proper pollinator. Other areas did not have the proper insect vectors. In the pitayas' native range the many types of Hylocereus produce spectacular fruits weighing up to a kilo in colors ranging from deep purple to pink. Large scales on the thin, leathery skin give an artichoke like appearance but inside is a solid ball of delicious edible fruit. The flesh is smooth and creamy and the taste like a combination of an excellent kiwi and a good melon. The plants are salt tolerant and drought tolerant though they are sensitive to frost and excessive sunlight.

The fruits are a good source of Vitamin C, minerals and sugars.

Attempts have been made in recent years to select and stabilize Hylocereus types. The fruit is being produced commercially in Nicaragua, Costa Rica, and Columbia. There has been some success in cultivating it in commercially Israel and Vietnam. While it was imported into Israel in the 1980's as part of a program to find new fruits for arid and saline areas; the Vietnamese horticulturalists claim it as their own often marketing it as the "national fruit of Vietnam" an amusing situation since all Hylocereus cacti originally come from the new world. However Hylocereus undatus, ("dragon fruit") has adapted so well to some areas of Vietnam that the mistake is a natural one.

Local Importance: Gaining in importance as a cash crop and export crop, grows on other plants in intercrop situations.

Potential Importance: Interesting new crop for arid and saline, frost-free areas.

Comments: Hylocereus varieties are actually different species. Very few of the wild types have been collected or cultivated.

Table 2: Red Pitaya—Food value per 100 gram of edible pulp

Calories	49-50
Moisture	85.4—87.0 g
Protein	0.13—0.17 g
Fat	0.7 g
Carbohydrates	10-11 g
Soluble sugars	5.9—6.8 g
Acids (citric)	0.1—0.9 g
Fiber	0.8—0.9 g
Ash	0.30-0.43 g
Magnesium	26.3—27.3 mg
Calcium	6.9—7.7 mg
Phosphorous	13.2—14.3 mg
Iron	0.52—2.80 mg
Sodium	12.0—12.4 mg
Potassium	206—242 mg
Vitamin C	6.6—7.3 mg

analysis by Ben-Gurion University

Argania: The Oil Tree of the Atlas Mountains

Photo by Elaine Solowey

Figure 3: Argania

High in the Atlas and Anti-Atlas mountains and in a narrow band of Moroccan coast the natural stands of Argania trees are in decline, eaten by goats, displaced by attempts to introduce modern crops to these remote areas (colorplate 9 on page 60).

Argania (*Argania spinosa*) became a "lost crop" in the 20th century.

The argania trees have grown here for thousands of years and existed as a semi-cultivated, semi-domesticated part of the landscape. The ancient Jewish pre-Islamic communities of the Atlas mountains mentioned the argania oil's health-giving properties in their writings. Argania oil was an important trade item from this area from the 12th century to the 19th century brought by Arab traders to the markets of Tangiers, Cairo, Damascus and Baghdad.

It was there that French colonials became acquainted with the properties of the golden orange oil with the nutty taste. Soon argania oil was being used in gourmet cooking and incorporated into cosmetics and perfumes. In the 20th century the argania went into its precipitous decline, the total stands reduced by 50 percent in as many years.

Part of the reason for argania's disappearance is its value as a forage crop for goats and camels that relish the thorny foliage and small juicy leaves. Small trees are often coppiced to provide tough poles and larger trees cut for their weather-resistant lumber (The nickname of the argania tree is ironwood). In its native range the argania tree's yield of nuts is often very low making other uses for the land look more profitable.

The argania produces two falls of nuts per year in areas with over 300 mm of rainfall and one fall a year in dryer areas. A deep extensive root system and the ability to shed leaves and survive as a dry bundle of thorny stalks has given the argania a natural advantage in the driest zones of the Atlas and anti-Atlas where it is often the only tree in the landscape. Argania trees can endure drought-induced dormancy for many years then put up green shoots quickly when the rains return.

The nuts look much like acorns. When they are cracked the heavy shells can be separated from the almond like kernels and the oil extracted by cold pressing. The oil is a deep orange color and smells very much like walnut oil. It is not suitable for frying but as a healthful salad oil it is without compare.

The oil can also be used in cosmetics, perfumes, and salves without further processing.

Traditional oil producing methods include feeding the encased fresh nuts to goats then picking the indigestible nuts out of the goat dung after the outer layer has been digested by the animals. The nuts are then washed, roasted, cracked by hand and pressed.

Mechanization of this process, leaving out the goats, produces superior oil and may make argania oil more acceptable as a commercial product.

Argania trees have been introduced successfully to Israel as a crop candidate for Israel's barren south. Trees have been introduced experimentally to the several other countries. Large-scale plantings in other north African countries have so far failed.

The tree is extremely drought tolerant and salt tolerant but is vulnerable to fusarium fungus and can be killed by temperatures under -8°C.

Local Importance: An ancient crop lost in the 20th century. Source of *amamlo*u nut butter. Local botanical treasure now in decline.

Potential Importance: Land reclamation, agro forestry tree. Source of edible oil for arid and saline areas. Timber producer, anti-erosive, subsistence tree for the most barren areas. Possible medicinal oil.

Comments: There is renewed interest in argania cultivation around the Mediterranean as fresh water becomes scarcer.

Amaranth: The Incomparable Staples of the Aztecs

Three species of amaranth, neglected for centuries, could substantially increase protein production in developing countries. Amaranths are fast growing cereal-like plants that bear small seeds in flower like heads. The seeds are extremely rich in protein and the leaves of the plant can be harvested like spinach over much of the year.

Amarathus hypochondriacus

The most robust of the grain amaranths was the major grain crop of the Aztec civilization and was violently suppressed by the Spanish in an attempt to eradicate "pagan" ceremonies which had developed around the cultivation and harvesting of amaranth. This type of amaranth survived as a minor crop used in making a delicious sweet called alegria, baked into festival cakes and popped into a light snack.

This species is very useful for tropical areas and high altitudes; its seed quality is excellent and shows the greatest potential as a food ingredient. This species of amaranth also mills well and has a pleasant taste and smell.

Seeds taken by the Spanish back to Europe were planted in private gardens and soon Amaranthus hypochondriacus was accepted as an ornamental plant and appeared in numerous illustrated herbals of the 16th and 17th century. By 1700 it was being cultivated as a minor grain crop in Central Europe and spread from there through Russia into Asia. Now it is grown as a hill crop in locales as widely separated as India, Ethiopia and Mongolia.

Oddly enough the amaranths of China and the Himalayas seem to be ancient well-established cultivars. Though these strains of amaranthus hypochondriachus are obviously of American origin certain genetic differences point to a very early date of introduction. This is something of a mystery as the presence of the plants indicates there might have been some kind of contact between Asia and Central America many years before the Common Era.

Local Importance: Crop relic, minor grain crop.

Global Importance: Protein production for the developing world, old/new, adaptable grain and vegetable crop.

Comments: Valuable in many formats.

Amaranthus cruentes

These Mexican and Guatemalan species are used both as a grain and a leafy vegetable (colorplate 10 on page 61). The grain varieties have white seeds and the vegetable strains tend to have dark, flavorful seeds, which are slightly harder to mill. It is the most adaptable of the amaranth species and shows an ability to thrive in a variety of day lengths.

Cruentes seeds have been found in burial caves undisturbed for 6,0000 years making it one of the earliest known domesticated American crops.

Amaranthus cruentes has survived as a crop relic in some villages of Mexico and Guatemala where it was usually popped for inclusion in confections.

In the American southwest it was used by pueblo Indians to color corn-based food.

Local Importance: Crop relic, local specialty.

Global Importance: Grain and greens for a wide range of temperatures and climates. Not sensitive to photoperiods.

Comments: Better methods of seed collection and cleaning would aid commercial cultivation.

Amaranthus cadatus

This plant is in the process of being "found", that is, recognized by mainstream agriculture, investigated and hopefully improved until it can take its place as a modern crop plant (colorplate 11 on page 61). Cadautus plants are exceptionally sturdy and prolific with large and palatable seeds. The leaves of the cadautus varieties are flavorful and very tender when young, often veined in red and yellow. These leaves can be eaten as a vegetable without cooking.

One strain of *Amaranthus cadautus* (its indigenous name is Kawhicha) is now being cultivated in twenty countries with a good deal of agricultural and economic success. It is the first of the grain amaranths to become a comeback crop probably because of its value as a health food and wheat substitute and its suitability for inclusion in baked goods.

Local Importance: High altitude staple.

Potential Importance: High protein comeback crop.

Comment: Amaranthus cadautus can be grown at very high altitudes. See "Kiwhicha: Grain Amaranth of the Andes" on page 109.

Many "lost crops" can be brought quickly back into cultivation.

Others are gone, though their undomesticated ancestors may continue to exist in the wild places on the fringes of the "lost crop's" native range.

Seeking out these valuable plants and restoring them to our agricultural systems is an important step on the path to true abundance.

Photo by Ruthi Rubenstein

Colorplate 1: Lucuma

Photo by Ruthi Rubenstein

Colorplate 2: Sapodilla

Photo by Professor Yossi Mizrachi

Colorplate 3: Cereus Fruit

Photo by Elaine Solowey

Colorplate 4: Cereus Fruit

Photo by Elaine Solowey

Colorplate 5: Sunflowers

Photo by Professor Yossi Mizrachi

Colorplate 6: Cereus Fruit

Photo by Ruthi Rubenstein

Colorplate 7: Pitaya

Photo by Elaine Solowey

Colorplate 8: Pitaya Harvest

Photo by Elaine Solowey

Colorplate 9: Argania

Photo by Elaine Solowey

Colorplate 10: Amaranth cruentes

Photo by Paula Keys

Colorplate 11: Amaranth cadautus

Photo by Scott Vlaun (reprinted from Seeds of Change)

Colorplate 12: Rainbow Inca Corn

Photo by Scott Vlaun (reprinted from Seeds of Change)

Colorplate 13: Bolivian Rainbow Chile Peppers

Photo by Scott Vlaun (reprinted from Seeds of Change)

Colorplate 14: Tepary Beans

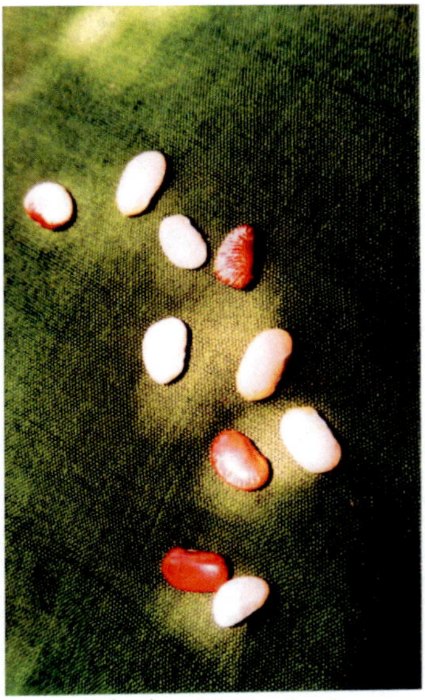

Photo by Ruthi Rubenstein

Colorplate 15: Popping Beans

Photo by Ruthi Rubenstein

Colorplate 16: Quinoa

Photo by Ruthi Rubenstein

Colorplate 17: Finger Millet

Few people realize that the food supply of humanity is tied to the genetic variability of crops and crop relatives. This genetic diversity has provided the raw material for plant domestication for the last 12,000 years. But in the 20th century much of this incalculable genetic wealth is simply being thrown away.

Older cultivated stocks, crop relatives, and wild crop ancestors are disappearing because of neglect, loss of habitat and the simple fact that agribusiness does not value the very stuff of life that has allowed modern crops to come into existence.

Modern crops are mostly strains that are profitable and useful today in a system that uses immense amounts of water, chemical fertilizer, fuel, pesticides and herbicides.

But What if Conditions Change?

If the globe becomes warmer—or colder? If the need to stop global warming forces us to rethink the way we use energy?

If diseases and pests mutate and attack crops that we believed to be resistant?

A brief glance at the history of agriculture shows us that conditions can change very quickly. In 1200 a long period of mild weather ended leaving Danish colonies in North America and Greenland isolated by icepacks. The colonists died of starvation, disease and hostile attacks by 1350. The Little Ice Age in Europe (1430-1850) caused widespread crop failure and starvation. A wet spell in the 1840s changed the course of Irish history and sent three million souls to other countries in search of food.

Where will the new crops come from when conditions change again? If we lose the landraces, heirloom varieties and wild ancestors that have allowed us to feed ourselves for so long, will the modern varieties and isolines be the end of agriculture as we know it?

Identical Twins: Identical Weaknesses. In 1970 a blight wiped out 15 percent of the American corn crop in less than six weeks. Only favorable weather patterns kept the losses from being much higher. Eighty percent of the corn was vulnerable because all those acres of plants carried an identical genetic sequence that made them susceptible to a new strain of blight.

This problem cannot be corrected by technology or by added use of pesticides and fungicides.

Corn plants in America are "as alike as identical twins." The genetic uniformity of these plants *is* the problem. (Raeburn, 1995, pg. 121)

The Last Harvest In Paul Raeburn's brilliant book *The Last Harvest* the danger to the world's food supply is elegantly and concisely explained.

Millions of acres of identical plants offer unprecedented opportunity for pests. If an insect, fungus, or bacteria mutates enough to attack one plant successfully, then it can attack all of the plants wherever they are grown.

In 1970 the corn blight spread with such astonishing rapidity that a thriving cornfield could be turned in less than a week into a nightmare of dry seared-looking stalks that crumbled into clouds of spores when touched.

The most famous incident of this type is still the "potato famine." In the 1840s the potato blight wiped out most of the potatoes in Ireland causing close to two million deaths by starvation. These potatoes were all descended from the same south American potato clones and had no resistance to the potato blight, which originated in Mexico. To make a bad situation worse, the Irish peasants had no alternative food crop as food grains grown in Ireland were part of the rent paid to absentee landlords.

In 1916 red wheat rust destroyed 40 percent of America's wheat. Striking again in 1954 a slightly different rust disease destroyed 75 percent of the wheat.

In the last twenty years citrus, grapes, rice, coffee, apples, pecans and peaches have been devastated by disease, usually because of the genetic uniformity of the crop in affected locales.

The only practical solution is through maintaining and cultivating biodiversity in crop species. This involves growing many different kinds of crop plants with diverse characteristics and diverse genetic makeups.

The Reverse Pyramid But modern agriculture is based on widespread monocultures, huge dense stands of billions of identical plants that cover thousands of acres.

Modern plant genetics is founded on the "pure-line" mentality, which devotes much energy to producing uniformity of size, taste, texture, harvest index, height and color in crop plants.

The genetic base of crop plants becomes increasingly narrow with more and more acres planted in fewer crops and fewer varieties.

For obvious reasons this mindset puts much of the world's food supply in danger.

Between 1900 and the year 2000 the US corn crop lost 95 percent of its genetic material. Older varieties of corn were simply not grown because of the higher yields of commercial hybrids. Native American agriculturalists who were still growing the older stocks were relocated, cheated of their land or coerced into adopting modern farming methods. Small farmers were forced out of business because the commercial hybrids were part of a very expensive package which included machinery, pesticides, herbicides and fertilizers that allowed only the biggest and more successful growers to compete. Later many of *these* farmers were eliminated by the rise of the corporate farm.

Ninety five percent of the corn was eliminated and forgotten as the people who had raised it were forced out of agriculture.

What was left was hybrids of hybrids and clones of clones.

The situation is not much better in other cultivated crops. In the US 80 percent of the vegetable varieties available in 1920 are now extinct. Australia has lost 2,245 useful plant strains in the last fifty years. South Africa has lost 2,215.

All these nations are in the developed world and all of them are allowing or encouraging the extinction of the very genes that will feed humanity tomorrow.

Our Stake In Plant Diversity John Tuxill of Worldwatch Institute has written an in depth report on the loss of biodiversity called *Nature's Cornicopia: Our Stake in Plant Diversity.*

Worldwide more than 30,000 plant species are threatened.

> *Even crops that may not be producing or seem undesirable could one day help fight against human disease, develop resistant varieties or respond to changing conditions. Although we have unprecedented skill in moving genes around only nature can manufacture them. If a plant bearing a unique genetic trait disappears there is no way to get it back. And you never know what is going to useful down the road. (Tuxill,1999, pg. 4)*

As an example Mr. Tuxill mentions a gene for resistance in high yield US wheat which was found in a wheat strain in Turkey which has since become extinct.

Advocates of biodiversity often say that letting an organism become extinct before we know its role in the ecosystem is like "burning a book before you read it."

To take this analogy a few steps further, if the crop plants of the US were like an entire library in 1900, now they have been reduced to a single shelf of books.

The commercial strains of many crops will not endure the inevitable collapse of industrial farming. They were impoverished and debilitated by the quest for uniformity. The crops are as over bred as the American white domestic turkey and the French poodle.

Where will the new crops come from?

The Idea of the Seed Bank Indigenous agriculturalists in the American Southwest were the inventors of the seed bank. The seeds of the most important food plants were stored in closed gourds in sealed caves, sometimes for as long as a decade.

Especially good seeds were also preserved in this manner to protect the best of the crop in case of disaster.

The indigenous farmers knew how important crop diversity was to their survival and preserved their cultivated strains with great care. Sometimes a single community would grow twenty different strains of corn, squash or beans. They also encouraged the growth of the wild relatives of their most precious crops, recognizing that there was something in the wild strains that was good for the domesticated ones.

Indigenous farmers did not discard any cultivated plant, knowing that under different conditions the weak crop might become strong and the strong crops could be weakened.

Vavilov, Seed Banks and the Centers of Crop Diversity

Nikolai Vavilov is the father of the *inexact science of gene prospecting*. (Raeburn, 1995, pg. 38) He is also the first of the great seed collectors. His interest in crop diversity was forged under fire in WWI. Vavilov was not drafted because of vision damaged in a chemistry experiment. However the Russian army asked him to help solve a mystery: local bread on the Iranian front was causing an epidemic of ergot-like symptoms in the soldiers who ate it. But the bread was made from wheat, not rye, which is not infested with the ergot fungus.

Vavilov was sent to northern Iran to investigate. There he discovered that the flour was contaminated by a poisonous weed. The authorities were pleased and let Vavilov explore the fields of Iran and return to Russia with a huge collection of wheat and rye.

By 1921 Vavilov had acquired government support and moved to St. Petersburg (later called Leningrad). There he became the head of the All Union Institute of Plant Industry. From the Institute he launched collection expeditions that took him all over the world. Before visiting a country he would learn the local language. At the end of his life he spoke twenty two languages and could quote Islamic and Hebrew religious texts in the original.

The plant institute employed twenty thousand researchers and kept the staff of four hundred laboratories busy all year round. Vavilov's staff spread the collected seeds across the Soviet Union, greatly enriching local crop diversity with resistant, unusual and cold-hardy strains of many cultivars.

In the course of his collecting Vavilov identified the centers of origin and crop diversity of many cultivated plants. In certain locales he would find a profusion of varieties where, according to his theories, the crops had had the most time to evolve and diversify.

Vavilov's botanical map of the world would later help greatly in the search for disease resistant genes to save and improve modern cultivated crops.

Vavilov's seed bank system became one of Russia's most valuable resources. Other researchers, realizing what a genetic gold mine Vavilov had discovered, scrambled to establish similar collections in the US, Britain, Germany and Sweden.

Yet Vavilov's collection almost did not survive WWII. Leningrad was besieged by German troops, food supplies failed and it became known to the starving citizens of the city that the seed bank was full of potatoes, peas, beans and other edibles. Security was tightened and the seed collection was kept under constant guard. The staff of the institute rightly believed that the collection would be needed after the war to help reestablish Russia's food crops.

Thousands died of starvation in Leningrad. Among them were several crop scientists from the institute who perished, surrounded by packages of seeds, rather than eat the precious samples in Vavilov's collection.

Vavilov himself died of hunger in a Soviet prison. Stalin had found the pseudo scientific theories of the notorious Lysenko more in tune with Soviet political ideas. Vavilov attacked Lysenko's ideas and denounced them as nonsense. Stalin had Vavilov imprisoned when he refused to recant.

This was a tragic, ironic end for one of the most visionary scientists in the history of mankind. But among Vavilov's lasting contributions was the idea of the modern seed bank as a repository for the seeds which might be needed in the future and so must be preserved.

Under Funded and Undervalued The protection of crop diversity should be one of the foremost agricultural endeavors since all future crops will be bred from what we are able to preserve now. Yet seed collections are chronically under funded and undervalued. The importance of the stored germplasm is dimly understood even within the agricultural departments and ministries of the world's governments. In *The Last Harvest* a disturbing number of official collections are referred to as "seed morgues" (Raeburn, 1995, pg. 64). Because of improper storage, careless handling and general lack of cash many specimens are lost and many cannot be regrown.

Rip Van Winkles of the Plant World Rip Van Winkle, so the old tale goes, sleeps for forty years. When this legendary character wakes from his sleep and tries to go home everything has changed. His wife is dead, his children are grown, his house has tumbled down and no one remembers him.

Plant species kept in a refrigerated vault are like Rip Van Winkle. The world has gone on without them. New diseases have arisen, the genetic make-up of the pests has changed and the protected plant has not.

Resistance is only forged in the presence of parasites as many plant breeders have discovered to their chagrin. When a plant species is removed from the evolutionary forces that shape it, it becomes just what most modern crop plants are: static, defenseless, sitting ducks for diseases and pests to infest, dependent on crop protective chemicals to survive.

Gary Nabhan, recipient of a MacArthur foundation "genius grant," founder of the Native Seed Search and author of the book about native American agriculture called *Enduring Seeds* is a fierce advocate of in situ preservation of plant diversity. He gives this example: *Let us imagine that a crop breeder searching for salt tolerant plants discovers that a twentieth century scientist once banked seeds of a wild bean relative that grew on the saline playas of Mexico. That wild bean population is now extinct and no other sources of salinity tolerance are known for cultivated beans.* (Nabhan,1984, pg. 97)

Nabhan's parable does not end happily. The breeder grows the beans in his greenhouse and uses biotechnological tricks to pass the salt tolerance to cultivated beans. But the plants become infested with a virus that did not exist when the seeds were collected by the seed bank.

This parable of the future bears an eerie resemblance to what actually happened in the rice crop in the early 1970s.

Most of the traditional varieties of rice had already been replaced by IR8, a "Green Revolution" strain, and later, by IR20, an improved dwarf strain.

These two high-yielding strains were not particularly resistant to disease.

IR8 did not always produce crops of sufficient quality and was attacked by bacterial blights.

IR20 was vulnerable to the grassy stunt virus and the brown plant hopper.

A mighty breeding effort brought forth IR36 which was both of good quality and resistant to most pests.

In the ancestry of this super rice is a wild strain collected in India called *Oryza nivara.*

Of the six thousand samples of rice at the International Rice Research Institute this rare wild rice was the only sample resistant to grassy stunt virus. The rare and apparently worthless wild rice strain contained the genetic sequences which allowed the new rice to survive.

This time the story ended happily.

What about the challenges of tomorrow?

Treasures in the Backyard The following common plants have uncommon features in their immense, uncataloged and largely unappreciated genetic diversity which can help us achieve true abundance.

Maize: The Divine Gift to the New World

Photo by Kuthi Kubenstein

Figure 4: Corn

Maize (*Zea mays*) may be the most important crop on the face of the earth (colorplate 12 on page 61). More wheat and rice is eaten directly but maize underpins most animal feeding programs and has so many uses in industry and medicine that it is probably not possible to list them all. It is also not possible to name all the food items which contain corn syrup, corn starch or corn oil.

Maize is also a vast plant family, with three main races, fifty sub-races and thousands of strains, ecotypes and varieties.

Maize grows in the tropics along with such useful tropical plants as sugar cane, it has been planted in vast belts in temperate zones and it ekes out a useful existence in chilly uplands where its kernels are eaten by man, its cobs by animals and its dry stalks become fuel after the short growing season.

Maize includes the spindly teosintes which some researchers believe are the ancestors of modern corn (and other believe are a weedy *descendent*), popcorn, podcorn, huskcorn, flour corn, dent, flint, blue corn, milk corn, sugar corn, rice corn and stone corn. There seems to be a form of maize for all uses and all seasons.

While 95 percent of corn's genetic wealth is gone from the United States and 80 percent of the genetic material has been lost in Mexico there still remains an immense treasure of corn races and types from South America to China which must be preserved if maize is to retain its vigor and adaptability

Triumphant Tropical Maize Maize has incredible rebound ability, that is the ability to marshal its own defenses to adapt to new conditions or overcome disease or pests.

An illustration of this strength is given by Raoul A. Robinson in his important book *Return to Resistance.*

Maize was taken to Spain shortly after the discovery of the New World by Columbus, probably from the islands of the Caribbean. Maize adapted to the cooler and drier Spanish climate. After approximately fifty years maize was brought to the Canary Islands and then to tropical Africa. There the maize developed into a group of landraces and cultivars that came to be known as Tropical Maize. In its travels however it had left the parasites of the Caribbean far behind.

In the 1940s diseased cobs from the Caribbean were accidentally introduced to tropical Africa. A species of tropical rust, a virulent reencounter parasite, devastated the tropical corn crop. Plant breeders and researchers scrambled to find resistant strains but as soon as they thought they found a source of transferable resistance the strain fell prey to the swiftly mutating rust organism.

Meanwhile the farmers who had grown tropical corn for so long continued to plant the stunted survivors of the rust. Over a period of a decade the dormant and diffused genes for rust resistance accumulated in the surviving corn and rust became a minor pest again instead of a lethal epidemic.

The scientific plant breeders never did find a solution to the problem. (Robinson, 1996, p 173-199)

Other Strengths of Maize Genes for salt tolerance, cold tolerance, enhanced starch, enhanced sugars even genes which could turn maize into a perennial plant, all exist within current maize types.

To use the book analogy at the beginning of the chapter, if the genome of a plant is like a book, then maize has opened a library on each continent and these libraries on the most part are still full of unread material. This is no reason to be complacent because we do not know what we will need in the future to make sure that this wonderful plant continues to feed mankind.

Mysterious Maize

Maize is not like any other existing plant and its wild relatives are so far removed that it is not possible to tell if maize somehow developed from one of them or if they are the result of some strange hybridization between maize and a species of grass. The upright stem of a maize plant is called the *stalk*. Side branches are called *tillers*. At the top of the stalk is the male flower or *tassel*. The female flower eventually becomes an *ear* when the pollen falls on the *silks* which are actually styles leading back to a single kernel. Looking at maize with an objective eye one wonders how such a plant could have developed in nature. Indigenous farmers who grew maize and depended on it for their bread may have lived anywhere from the American Southwest to the uplands of the Incas but they agreed about one thing: maize was a divine gift to mankind, not a product of human skill with plants.

Garlic: Both Food and Medicine.

A very old Jewish folk tale tells of a man who lives on an island blessed with a long agricultural history and a wealth of cultivated plants. He travels to the mainland and discovers that garlic, a common garden plant on his island, is completely unknown. Returning to his own nation he fills a ship with garlic and sails triumphantly to the mainland where he presents his shipload of garlic to the king.

The king is mightily pleased with the new plant which can be used to flavor food and preserves human health as well.

"You will be rewarded with the best my land can offer!" the king tells the islander and he

fills the islander's ship with incense, apes, peacocks, silver and gold. The king distributes the garlic to his most favored subjects to cultivate and garlic becomes so valuable that a silver piece will buy just one bulb of it in the marketplace.

The islander returns home, a rich man.

Inspired by the story, the islander's neighbor fills a ship with onions and sets sail for the mainland the very next year.

The king is delighted at yet another new useful plant from a far off place and makes much of the neighbor. "You will be rewarded with the most valuable product of our land," the king decrees.

The king then proceeds to have the neighbor's ship loaded with garlic.

Garlic (*Allium sativum*) is one of the older cultivated plants. It grows easily in most soils and climates and has consequently found its way into every national cuisine. Wild garlics are relished in even the most isolated human groups. Where did this plant come from? What can we learn from something as common as garlic?

Chinese garlic cultivation traditionally dates back to the Emperor Huang-ti of the first dynasty. He journeyed up a mountain with his followers and was accidentally poisoned when nonedible plants were added to the evening meal. Luckily wild garlic grew on this mountain in abundance. The emperor and his followers were cured by a hastily concocted garlic remedy. This led the emperor to promote garlic's cultivation in China.

Garlic was also known to the ancient Hebrews, who lamented all through the Exodus that they had no garlic to flavor their food. It is unlikely that Chinese garlic and the garlic they longed for originated in the same area.

Interesting enough slaves and laborers in Egypt were given daily rations of garlic to help them keep up their strength. A strong male slave could be bought for 15 measures of garlic.

Six cloves of garlic were found in King Tut's tomb. To flavor the Pharaoh's food in the afterlife? Or protect him from harm?

There is some evidence to suggest that the Romans brought garlic back from the Middle East and use of garlic spread through Europe with the Roman occupation. But wild garlic was already eaten in England 4,000 years ago even though the Normans are credited with introducing cultivated garlic from Europe.

What makes all this so unusual is that garlic is a plant that has for the most part lost its ability to set seeds. Garlic bulbs must have passed from hand to hand as trade goods or perhaps as a form of currency since most of the cultivated garlic of our day is descended from a limited stock of veteran clones. Perhaps there is something in the smell, flavor or appearance of garlic that speaks to our deeper instincts and tells us that garlic is a desirable food.

Ancient Clones

These garlic clones are of ancient origin. Despite the fact that most plant clones quickly become targets for disease because they are genetically static, garlic shows no loss of vigor even though the clones are thousands of years old.

The importance of garlic in preindustrial age diets has been greatly underestimated. Garlic was both food and medicine and probably saved more lives than the crude conventional drugs of the day.

Modern researchers have discovered antibacterial, antifungal and antibiotic properties in this extraordinary plant and continue to discover even more interesting properties.

Questions that should be asked about garlic: Is there any way to return garlic to the increasingly unhealthy diet of the industrial West?

Is there a lesson for modern agriculture in the continued vigor of ancient garlic clones?

Consumption of garlic seems to have many beneficial health effects in the developing world, even among the chronically malnourished. Is there a place for garlic in the feed-the-world strategies of the next century?

Should the wild ancestors and relatives of cultivated garlic be investigated for their genetic characteristics and medicinal properties?

Peppers: Full of Nutrition and Flavor

An old South American tale tells of an Inca prince out walking in the jungle. Three spirits who wish to try life as humans plan to accost him. One spirit means the young man no harm but would like to become his wife. The other two intend to become human by possessing his body.

While he is resting by a pool of water each of the three spirits speak to him. One speaks from the jungle orchid, one from the fever tree and the last from a pepper plant growing by the pool. He is told he must choose one spirit to accompany him or he will not be allowed to leave.

After much thought the young man chooses the spirit in the pepper, thinking that no evil could dwell in such a homey, useful plant When he makes his choice the spirit in the pepper plant is transformed into a handsome young woman who goes home with him as his bride.

Like potatoes, peppers (*Capsicum sp.*) were universally accepted and spread quickly from South America to the entire globe (colorplate 13 on page 61). Also like potatoes, peppers are one of the few vegetables that are still found in a wild state.

In a third similarity to potatoes, peppers have a vast range of size, color, flavor and genetic variation that is mostly untapped.

Colin Tudge in his book *Food Crops of the Future* mentions the importance of tertiary foods (herbs, spices and condiments) in a proposed diet for a planet of ten billion. *If the projected world population of 10 billion is to survive then people must in general subsist on a simple diet—one with a high proportion of staples. But such diets are not tolerable for long unless they are made interesting and herbs and spices can make them so.* (Tudge,1988, pg. 38)

The people who must subsist largely on rice, millet or potatoes do not have to be told this. Paul Richards who wrote the book *The Indigenous Agricultural Revolution* notes that extremely poor people in Africa often spend their last few coppers on spices to make their diet more palatable. He comments that this can be a wise investment. (Richards, 1985, pg. 25)

The pepper is both spice and vegetable, brightly colored, flavorful, full of vitamins and antibiotic properties. It is also a stimulant to flagging appetites and a rich source of Vitamin C.

Unlike many spices it can be grown just about anywhere.

It is believed that peppers originated in Mexico and Central America but were taken into South America and up into the American southwest by indigenous farmers and traders.

In this time period the pepper developed into several dozen types and thousands of varieties.

After the Europeans reached the New World the pepper quickly spread to Europe. It was adopted so swiftly in the East that many Indians and Chinese will swear that the pepper has been part of the national cuisines since time immemorial.

The pepper family ranges in taste from the very sweet to the extremely biting; in color from pale yellow through all shades of red and green to shiny black. Peppers are cultivated in Israel, in Australia, in tropical Africa and Nepal demonstrating their adaptability and usefulness.

What might we expect from the pepper in the future?

Firstly, it is an affordable tertiary food, one that turns a dull bowl of rice into a warming and nutritious main dish. Even the poorest people can grow a few pepper plants and a few plants are often all that is needed.

Secondly, the pepper family has an immense wealth of variation and characteristic. Roughly divided into sweet and hot types peppers can be further developed as both spice and vegetable. There are also numerous wild strains that should be investigated and preserved to invigorate cultivated stocks and perhaps create new pepper varieties.

Lastly, peppers are related to tomatoes, the most popular of "vegetables" (both tomatoes and peppers are actually fruits). Tomatoes are big business and increasing amounts of land and water are devoted to cultivating them. If tomatoes must be modified and manipulated for agricultural purposes then perhaps the needed genes can be found in the incredibly diverse pepper family instead of being culled from flounders, worms, bacteria and tobacco plants.

What are Heirloom Varieties?

Heirloom varieties are varieties of plants that have specific valuable characteristics and ecotypical attributes that made them especially important to generations of gardeners and farmers. Like family heirlooms they were passed from generation to the next. In the drive for standardization of produce that has taken place in the last fifty years many heirloom varieties have been lost or forgotten and all have been neglected by commercial and scientific plant breeders despite the fact that they are often hardier, tastier more useful and more resistant than modern varieties.

Sunflowers: Billion Dollar Blossoms

It was a bad dry year. The sun was twice as big and three times as hot as usual. Long strands of fire swept out of it scorching the earth. In order to help the suffering people Coyote leapt into the sky and struck the sun with a great club. The sun drew back, sparks fell over the hills and plains and where they fell the first sunflowers grew. Because the light of the sun is in their petals the sunflower petals are used in sacred ceremonies. The flowers know they come from the sun and that is why they turn their faces to follow the sun as it passes. From a story by Native American story teller Bill Simpson, Yosemite National Park, August 1967.

Sunflowers (colorplate 5 on page 60) are an oil crop with an estimated 250 million a year value in the US alone. The world-wide annual value easily tops a billion dollars. But in the United States ninety per cent of commercial hybrids share much of the same genetic material. This sets the scene for a disaster similar to the blight that struck the corn crop in the 1970s.

With such a high rate of uniformity the entire crop could be lost. There are over two million acres planted in sunflowers every year in the US.

Unlike corn, sunflowers originated in the United States and the centers of biodiversity are in the American Southwest. There one fourth of the native sunflower species are endangered and several have disappeared. But over half of the indigenous wild and semi-domesticated sunflowers still remain and steps must be taken to investigate and save them.

The high survival rate is not due to any protective measures on the part of the authorities but to the strength and quick growth habit of the flowers and the efforts of various grass root seed-saving societies.

Now is the time to preserve and catalogue the remaining genetic diversity of this important oil and energy crop. These wild crop relatives and older cultivars contain the genes for salt tolerance, drought tolerance, disease resistance and increased oil production that will be so valuable in the future.

Gene transfers from wild sunflowers to domesticated ones have already proved to be crucial to the sunflower breeding program.

Nine thousand years ago a sunflower relative called giant sumpweed (*Iva annua*) was cultivated as a source of oil. With the coming of the Europeans this intriguing plant fell out of use and became extinct in the space of a few years.

It is impossible to know what valuable traits could have been transferred from *Iva annua* to sunflowers or if giant sumpweed could have been further developed as a cultivated crop. It is simply gone and North American agriculture is poorer for it.

But it is not yet too late for the sunflower.

Jerusalem Artichokes: Neither/Nor

In the words of Jane Grigson, British food writer: Jerusalem artichokes are none the worse for not being artichokes and having nothing at all to do with Jerusalem.

Jerusalem artichokes (*Helianthus tuberosus*) are actually relatives of the sunflower. Whatever they are called they have a venerable history as cultivated plants having been domesticated approximately ten thousand years ago. Native American farmers planted the tall flowers as windbreaks to protect tender plants and to provide trellis for bean plants. At the end of the season the knobby storable tubers were harvested and eaten over the winter.

In 1616 the tubers were taken to Europe and called Italian girasole (turning sun) by some and Canadian artichauts by others. These two names blended into the odd misnomer "Jerusalem artichoke" and that is what the plant is called today.

Jerusalem artichokes became something of a forgotten crop for many years and much of their original diversity was lost. Still they survived in private gardens, in the plantings of Native Americans and in many areas in Europe where they took the fancy of the local population and were cultivated as ornamentals or minor vegetable crops.

A new appreciation of vegetable foods in the 1980s led to their return to commercial cultivation in California and a new name "sunchokes."

Sunchokes could become an important tuber crop. Their carbohydrates are not converted to sugar in the human body, making them a good food for diabetics and people who are watching their weight. The plants are vigorous and grow quickly with few inputs. They have several uses in intercrops and permaculture and some of their lost diversity might be compensated for by gene transfers from sunflower stocks.

A little care and minimum effort could save the unique characteristics of this "ordinary" plant for future generations.

Uncommon Bean: Unappreciated Jackpot

Beans (*Phaselous vulgaris*) are native to the new world and are the traditional complement to corn (colorplate 14 on page 62). Together the two plants give the human body complete protein. Beans are rich in minerals and contain soluble fiber which is so beneficial in the human diet. Beans come in a wide variety of tastes, textures, colors and habits of cultivation. They are hardy enough to grow in extreme climates and marginal lands

Beans are also an underestimated and neglected crop with a serious genetic diversity problem because beans have been relegated to the class of "poor people's food" in the last fifty years especially in the United States, which means that little money will be available to investigate, collect or preserve bean varieties.

Why a species of plants which have all the "jackpot traits" sought by scientific agriculture should be demoted to fifth class citizenship in the plant world is almost beyond understanding.

The "jackpot traits" are salt tolerance, drought tolerance, pest resistance and nitrogen fixing ability.

Some beans are so salt tolerant that they can be grown on the shores of the Dead Sea.

There are bean varieties that can grow to maturity after being watered *once.*

Most bean cultivars carry genes for "field resistance" or tolerance of diseases and insect pests, some to a very high degree.

All beans have nitrogen fixing ability and so improve the soil in which they are grown besides providing nutrients for their own growth.

Beans are one of the few protein foods that can be grown in both the arid and wet zones of the tropics, areas in which it is difficult to produce enough protein of any kind. Many inhabitants of these zones are significantly underfed and the common bean is the only plant between them serious illness or death.

Vanishing Genes Despite the bean's global importance North American commercial bean varieties have vanished with unbelievable rapidity until five strains of hard beans account for 98 percent of the North American crop. String beans and snap beans still retain some diversity but this is also disappearing in the food industry's quest for uniformity.

For genetic diversity in North American beans one must go to an association like the Seed Savers or the Native Seed Search.

These organizations cannot take the place of a comprehensive program to save crop plant biodiversity. Until such a program begins they are stopping some of the worst of the genetic erosion by collecting, cataloguing, growing and distributing vanishing cultivars.

The Seed Saver's work with the cultivars of the American farming tradition has resulted in the rescue of thousands of crop strains and wild relatives.

The Native Seed Search has been successful in salvaging some of the unappreciated genetic riches of the Native American farmer.

They are only two of dozens of organizations that have sprung up around the world to halt the terrifyingly swift loss of our plant resources.

A 2003 catalogue of the Native Seed Search offers beans for cooking, keeping and eating green; beans for low elevations and high;

long season and short season beans; beans that are tolerant of heat, salt and cold.

There is a separate list for tepery beans and yet another list offering runner beans, lima beans and wild beans.

Much crop diversity remains in the rest of the world where the bean is still considered an important food.

But the centers of crop diversity for the bean are in Mexico, the Southwest US and Central America. The bean's strength and diversity should be preserved at the source.

A researcher at the National Academy of Sciences in Washington once said: *If you were stranded on a desert island and could have only four plants with you—-those four should be corn, beans, peppers and garlic. That way you will be well fed, you will have an appetite for your food and you will be healthy enough to enjoy it.*

That is basically a description of what food security would be like for the one and a half billion of our brothers and sisters who are chronically hungry.

To achieve this we must stop the careless squandering of precious genetic resources.

And learn to appreciate the treasure in our backyards.

The Apples of El Camino Real

Fruit specialist Ron Walser and horticulturalist Estaven Arellano are looking for apple trees that date back to Spanish colonial times. These trees were planted on land grants in the 1700's along the Camino Real, the official King's highway that ran from Santa Fe to central Mexico.

The idea is to create a colonial heritage orchard, a kind of genetic library of living trees that will preserve the heirloom apples for future use and future generations.

So far 15 different varieties have been identified that "have their roots in Spain."

In addition to the colonial trees there are many trees which were brought to New Mexico in the 1800's including Starking delicious trees that originated in Iowa. Baldwin apples from Massachusetts and Grimes golden apples from West Virginia. These varieties came west with American pioneers.

Fruit growers have become interested in old hardy apple varieties. A good niche market is starting to develop for heirloom apples and these tough time-tested trees are often more suitable for new organic and sustainable plantings than newer strains.

Walser and Arellano are both working with the NMSU Sustainable Agricultural Research Center at Alcalde on this project and on other projects to preserve the botanical treasures in New Mexico's backyards.

VI. Ancient Grains

There are many good reasons to turn our attention to the breeding, cultivation and improvement of ancient grains. Grain is the most important food for humans and domestic animals accounting for 70 percent of all food energy consumed. Unfortunately six grains account for over 95 percent of the grains bought, sold and traded on the world market. These are called the **"Big Three,"** corn, wheat and rice; and the **"Little Three,"** oats, barley and rye. Human survival therefore depends on a remarkably narrow band of genetic material found in six species of plants that have limited ranges of cultivation.

This is a precarious situation and must be remedied if humanity is ever to have food security.

Fortunately there are several ancient grains and wild grains that could be just as valuable to the world as the Big Three and the Little Three.

These ancient grains can be eaten like modern cereal crops or take the place of cereal crops as animal feed. A goodly percentage of wheat, corn and rice are lost to the hungry populations of Asia and Africa because it is being fed to the meat and dairy animals of the developed world.

These ancient grains can be grown in areas where harsh conditions make the cultivation of ordinary grains out of the question; areas too hot, cold, swampy, stony or alkaline. Some are nutritionally superior to conventional cereals and all use relatively small amounts of resources and water.

Also important is the great genetic diversity of these ancient landraces. These new fresh genes are needed in agriculture as the genetic base of our most important cultivars weakens and narrows.

In short, here is a way to diversify and extend the supply of cereals with little effort on the part of the agricultural and scientific community

Such an opportunity cannot be ignored in an increasingly hungry world.

Quinoa: An Ancient Crop Makes Good in the Modern World

Quinoa (*Chenopodium quinoa*) was the staple grain of the Incas considered so sacred that the Inca emperor broke the soil with a golden spade to plant the first seeds of the year (colorplate 16 on page 62). It is still used by millions in the highlands of Bolivia, Ecuador, Chile and Peru.

As a food it is unusually nutritious, containing 58 percent starch, five percent sugar, 12-20 per cent proteins and four percent fat. The protein quality is so high it can take the place of meat in the diet. The essential amino acids of lysine, methionine and cystine are present in quinoa in contrast to most other grains that are deficient in lysine and beans that are lacking in cystine and methionine.

Until recently quinoa was a classic example of a "lost" crop: bitter saponins had to be leached from the grains' outer layers, it was considered food for animals and poor people who could not afford to buy conventional grain. It was unknown outside of its native range and ignored by agricultural scientists and the food industry.

All this began to change in the 1960s. Quinoa attracted the attention of researchers from the National Academy of Sciences because of its hardiness in cool climates and high altitudes and the nutritional value of its sorghum like clusters of grain. A saponin free variety that did not have to be leached to palatability was developed in Bolivia and the first serious attempts at growing quinoa in the United States were conducted.

Commercial milling techniques added to the grains shelf life and marketability.

By 1980 quinoa could be found in many health food stores and less then a decade later quinoa was being used in the manufacture of bread, cake, trail food and breakfast foods.

Now quinoa is being cultivated in twenty nations as a crop for areas with short days, low temperatures and uncertain water supplies.

It has also become an important item for the increasing numbers of people who are developing sensitivities to wheat and corn.

Local Importance: Still a major crop in the South American highlands, important subsistence food and major source of protein.

Global Importance: Increasingly important in the west as health food and high protein additives. An important candidate crop for marginal areas at high altitudes in India and China.

African Rice: Hardy Red Rice of the Swamps

African rice (*Oryza glaberrina*) a little known family of rice has been cultivated in West Africa for over 1,500 years. It is usually red in color, has a special aroma and flavor and is often nutritionally superior to Asian rice. Many distinct and useful types of African rice have been developed in those fifteen centuries including varieties that tolerate salinity, fluctuating water supplies, harsh climatic conditions and excessive iron. African rice grows quickly to choke out weeds and is resistant to diseases and pests. Despite all these intriguing qualities African rice is almost unknown outside of West Africa.

Some strains already equal the productivity of Asian rice. Though valuable in its own right African rice is close enough genetically to hybridize with Asian varieties and cross-breeding may improve the yields for both. Genes for tolerance of heat, salinity, and excess iron as well as resistance to diseases and pests may be transferred to Asian varieties.

Local Importance: An important "hunger breaker" or intra-seasonal crop which matures quickly and produces yields under adverse conditions. A preferred indigenous food for the people of West Africa with higher nutritive values than the commercial strains which threaten to replace it.

Global Importance: African rice can be grown in a broader range of soils and climates than Asian rice. Genes from African rice can be used strengthen Asian varieties.

Pearl Millet: Food when all Else Fails

Pearl millet (*Pennisetum glaucum*) has been called a "plant breeder's dream" (BOSTID, 1996,112). It is so robust and needs so little space that a plant can mature and set seed in a five-centimeter pot. Pearl millet matures so quickly that four generations can be grown in one year. It produces masses of seeds. One inflorescence can yield a thousand seeds. One plant can produce over 20 inflorescences. The current landraces include robust high protein types, aromatic varieties and strains so sweet they can be harvested in the "milk" stage, eaten like corn on the cob or made into syrup.

Pearl millet is also incredibly hardy, germinating at high temperatures in crusted soils, tolerating sand damage, wet-season mildew, insects and resisting the noxious striga parasitic weed which is one of the banes of African agriculture.

Although most pearl millets grow best as dry land cereals some varieties are found in Africa's wet and humid zones. These plants are so different in appearance from dry land millets that they were first thought to be different species. Only genetic testing revealed that the larger, rounder, whiter, quick ripening rainforest plants are closely related to the millet of drier climes.

Pearl millet has been grown in the United States for some time as animal feed. It can fully replace maize in poultry feed for example and shows great promise as a whole plant fodder for cattle. Its low requirements for water and nitrogen and tolerance of drought and acid soils could make it a crop candidate for areas of the Middle West that must be returned to dry land cultivation.

Local Importance: Famine breaker, staple and bridge crop in much of Africa.

Global Importance: Grain and fodder crop for areas returning to dry land farming strategies; drought tolerant, pest resistant booster crop; replacement crop for maize.

Finger Millet: Concentrated Energy

Finger millet (*Eleusine coracana*) is a crop upon which millions of people depend (colorplate 17 on page 62). It has been neglected by the scientific and international communities because it is a "poor person's crop," consumed only a few miles from the areas in which it is grown and rarely reaching the international grain markets. Consequently it is almost unknown outside of certain parts of Africa and India.

Finger millet is the most nutritious of the cereal crops, having more food energy than corn, wheat and rice and being rich in amino acids that are absent in more popular cereals.

This crop has many other advantages. It has adapted to a wide variety of soils, climates and conditions. Finger millet seeds can be stored for years without insect damage and it can be made into a cornucopia of foods: porridge, beer, bread, brandy, malted products and snacks similar to popcorn. The straw has value as fodder containing from 50-61 percent digestible nutrients.

The crop's major disadvantage is in the size of the seed, which is very small with the bran tightly bound to the endosperm. Wetting the grain to loosen the bran and high-speed mechanical milling has produced excellent flours and semolina-like products.

Local Importance: Finger millet is the staff of life for people who must live on starchy foods giving them three amino acids that are absent in other cereals (tryptophan, methionine, and cystine), 14 per cent protein and generous amounts of trace elements. Some evidence suggests that finger millet allows many Africans to survive and continue to work and learn on one meal a day, surely a valuable quality in a hungry world.

To quote the National Academy of Sciences, *Lost Crops of Africa*, Volume I *Outsiders have long marveled that the people in Uganda and Southern Sudan could develop such strapping physiques and work as hard as they do on just one meal a day. Finger millet seems to be the main reason.* (BOSTID, 1996, pg. 39)

Global Importance: Finger millet may be the answer to Africa's chronic famine and to easing food shortages in other rugged areas such as Nepal and the mountains of China. Finger millet is also a powerful anti-erosive crop and particularly good at binding the thin soils of hilly areas.

Fonio: The Famine Breaker

Fonio (*Digitaria exilis/iburrua*) is one of nature's "fast foods" being ready to harvest six to eight weeks after it is sown. It is excellent in taste, extremely nourishing and easy to grow on remarkably poor soil. Despite all these advantages fonio is also an unknown crop with almost no commercial value outside its native range in West Africa. The tiny rice like grain supplies an estimated 4 million people with an estimable quick growing food that thrives where other cereals fail. Fonio's combination of good taste and high food value could be of great future importance. Fonio contains more protein than eggs: the essential amino acids being methionine, leucine, isoleucine, trytophan, valine, threonine and phenyline. With mechanical milling the small size of the grain is not a disadvantage but to reach its full potential fonio probably should be probably sieved, milled and steamed before packaging as a commercial quality food.

The short brittle stalks are nourishing fodder for animals and fonio could be an important tool in short-term land reclamation projects.

Local Importance: Critical famine food for West Africa, esteemed for its excellent taste and suitability as an inter-crop.

Global Importance: Suitable for arid, marginal soils and depleted lands. A fast solution to famine in these areas.

Warning: Under certain condition fonio may be invasive. It is related to "crab grass."

Karamaka: The Super Sorghum of the Sudan

Sorghum (*Sorghum bicolor*) this not-so-distant relative of maize has just as much potential as its "Big Three" cousin. Indeed, its name means "little maize." Sorghum is the most biodiverse of the grain crops with a record 125,000 strains collected and classified. Besides the sorghum grown for syrup and animal feed there are popping sorghums, vegetable sorghums (eaten like sweet corn at the milk stage and resembling broccoli), beer sorghums, bird-resistant sorghums, parasite-resistant sorghums, bread sorghums, tannin free sorghums, cold tolerant sorghums and high protein sorghums. This immense mass of genetic material may be combined and recombined, hybridized and vybridized to produce some truly remarkable plants.

One such plant is the karamaka or "super sorghum" a vigorous, compact plant that looks very much like sugar cane that was discovered in 1984 by a team funded by the "Band Aid" concerts of that decade. An ecotypical cultivar of remote villages in drought-ravaged areas of Sudan, karamaka was found to have lysine content 62 percent above normal and a total sugar count of 35 percent. Its nutritional value was estimated to be two thirds that of milk protein. Karamaka's vigor is manifested shortly after germination with strong swift root development and multiple stems. Karamaka is immune to the striga weed, which infests and kills many African plants.

With no modern breeding program or use of irrigation or fertilizer karamka is comparable to the best of the modern cereals but is also drought resistant, pest resistant and suitable for poor, arid and saline lands.

Local Importance: Famine breaker, staple, subsistence crop.

Global Importance: A nutritional gem. A new cereal capable of helping feed the world.

Tef: The Mysterious Grain of the Ethiopian Highlands

Photo by Alison Brooks

Figure 5: Tef

Tef (*Eragrostis tef*) is a primary staple crop in one country in the world, Ethiopia. There it often supplies two thirds of the protein in the local diet and is planted more abundantly than any other cereal. There is nothing mysterious about tef's popularity in Ethiopia. It is a reliable cereal for an unreliable climate as it matures in 40 to 50 days and can be sowed late after a more delicate crop has used most of the moisture in the soil. It can also be sowed in drought years as it sets grain and produces straw for fodder under very difficult conditions.

Tef is a powerful anti-erosive, binding soil and establishing a network of roots in fifteen to twenty days undeterred by heavy soils, drying mud of flood plains or the baking heat and sudden cold of the highland dry season. It is the only cereal that thrives on vertisols, the notorious black cotton clays that sag, heave, crack and split due to fluctuating water supplies and unstable temperatures.

The tiny tef grains range in color from black to white. White seeds bring the best prices but the colored types have more robust flavors and as a rule come from hardier cultivars. Early varieties mature in 40 days, dry lands varieties in 60 to 120 days and late-maturing varieties in 120 to 160 days.

Tef does well in poor soil but responds favorably to moderate applications of manure. Tef can be harvested and threshed mechanically. The small seed size is not a problem when the proper equipment is used.

Without appropriate machinery tef requires a good deal of hand labor in both cultivation and harvesting.

Tef is more nutritious than wheat and contains no gluten. The small seeds cannot be separated easily from the bran so most tef is used as a whole grain product. Only tef can be used to make the famous Ethiopian staple injera, a spongy, pancake-like sourdough loaf that is the preferred food of most Ethiopians and has become something of a gourmet item worldwide because of its unique flavor and texture. Because of its mildly sweet, syrupy quality when cooked tef can be added to soups, muffins, cakes, pancakes, puddings, casseroles and gravies.

Tef has been grown for fodder in South Africa and has some success as a cereal crop in Yemen. In the United States it is now cultivated in commercial quantities for health food and gourmet food.

Local Importance: Staple and famine-breaker for Ethiopia, esteemed for its grain, soil conservation qualities and excellent straw and fodder.

Global Importance: Important crop candidate for highlands with poor and acidic soil and unreliable rainfall. Could be an important crop in Mexico, Bolivia, India, Pakistan, Nepal, Alaska, Canada, the former Soviet Union and northern China. Suitable for high altitude land reclamation.

Ancient Wheats: Einkorn, Emmer, Kamut and Spelt

These ancient wheat varieties are exciting for a number of reasons. They contain genes for valuable traits, which may be transferred to modern wheat varieties; they contain more protein and slightly different proteins than conventional wheats (some ancient wheat can be eaten by people allergic to the modern varieties); and they can be cultivated in areas where modern wheat simply cannot survive.

Einkorn

Einkorn *(Triticum monococcum)* was thought to be extinct. The earliest of all cultivated wheats fell out of favor and was replaced by more modern varieties because of the difficulty in separating the grain from the germ. Its superior flavor and its use in local specialty foods kept it under cultivation in parts of France and the southern Alps for thousands of years until its rediscovery in 1989 by botanist Jacques Barrau. The unground grain is used in making highly nutritious cakes and porridge. Its sweet, mealy buckwheat like kernels are very different from contemporary wheat. Einkorn is finding a market as a health food since it does not trigger allergic reactions in many people who are allergic to wheat.

Local Importance: Specialty crop, crop relic, cold weather food.

Global Importance: Contains useful traits, which may be transferred to modern wheats; is tolerate of cold and can be cultivated in rocky areas. Increasingly popular as health food.

Emmer

Emmer (*Triticum decoccum*) originated in the Middle East and was also thought extinct. In 1906, the original wild type was rediscovered growing wild on Mt. Hermon by the Jewish Palestinian botanist A. Aaronsen. Several daughter types of emmer have become useful crops, most notably in the highlands of Ethiopia, which has been named a center of crop diversity for 28-chromosome wheat. Duram wheat, which is used to make pasta, is an emmer type.

Wild emmer carries several traits, which might prove valuable in breeding disease resistant and drought-hardy wheats. Emmer is particularly resistant to wheat rust, which periodically attacks conventional wheat varieties. Other interesting traits found in wild emmer include early ripening, heavy kernels, dense spikes and short stems. Emmer wheats make excellent bread of great nutritive value with a good deal of natural sweetness.

This "fossil" wheat, unchanged from Biblical times is different enough from modern grain to be eaten by most wheat sensitive consumers.

Local Importance: Staple crop for the Ethiopian highlands, specialty crop for the developed world.

Global Importance: Carries traits, which can protect and invigorate modern bread wheats.

Kamut

Kamut (*Triticum dicoccum kamut*) which is cultivated for wheatgrass products, high protein flours and pasta is considered an emmer variety because of its unusual kernel size and high protein content. Thirty-six grains of kamut were supposedly discovered in an ancient tomb in Saqquar, Egypt, given to an American soldier and then passed on to the soldier's farmer father. The farmer then cultivated the grain and entered it in a state fair where its large kernels and buttery flavor attracted attention. The grain was then named King Tut's wheat.

This story may not be true but kamut is genetically very different from modern wheat and offers a non-allergenic alternative to modern wheat varieties.

Local Importance: King Tut's wheat has become the base for the "wheatgrass" industry. Wheatgrass sprouts can be eaten like alfalfa sprouts and the juiced young plants have become popular in health food circles in tonics, purifiers and nutrient cocktails.

Global Importance: Possible famine breaker, bridge crop and green food. Carries traits which can improve bread wheat and emmer.

Spelt

Spelt (*Triticum speltum*) was raised by the Stone Age inhabitants of Europe but abandoned as a crop for wheat types easier to grow and thresh. Spelt has long stalks compared to modern wheats and a husk that fits tightly to the kernel and is difficult to remove. This last quality which protects the kernel against fungi and insects is probably responsible for spelt's agricultural decline.

Spelt is making something of a comeback now because of its higher protein content and its greater concentration of vitamins and trace elements. It is one of the most nutritious of the wheats containing 14 to 17 percent protein while modern grain averages 11 percent.

Spelt's nutty flavor and resistance to pests have made it popular in environmentally sensitive areas and a special organization has been established to aid this stone-age wheat's comeback.

Local Importance: A comeback crop in Austria, Germany, Belgium and Switzerland which grows in cold and marginal areas.

Global Importance: Has valuable traits which can improve modern wheat's nutritional quality. Carries genes for cold resistance.

Kaniwa: A Grain for the Top of the World

Photo by Elaine Solowey

Figure 6: Kaniwa

Kaniwa (*Chenopodium pallidicaule*) grows well at altitudes where no other grain survives. In its native area temperatures rarely exceed 10°C and frost occurs nine months of the year. Even barley and quinoa cannot produce under these conditions. Consequently kaniwa is the subsistence farmer's safety net, sown over cultivated areas after tubers have been harvested and then left to fend for itself in the harsh climate until harvest. Kaniwa often reseeds itself like a weed and provides forage where no other forage plants grow. It is not a true cereal but a plant in the family of quinoa. The seeds can be eaten without leaching because of the absence of saponins in the husks. Kaniwa's amino acid balance is exceptional, being rich in lysine isoleucine and trytophan. It contains vegetable oil and carbohydrates in generous measures.

Local Importance: High altitude subsistence food and famine breaker.

Global Importance: Genes for improvement of quinoa, a new crop for extreme highlands and other cold arid areas.

Kiwhicha: Grain Amaranth of the Andes

Kawicha is a subspecies of the most interesting (*Amaranthus caudatus kawicha*) of the amaranths. Amaranthus cadautus was the staple grain of the Incas, Aztecs and other pre-Columbian peoples and was as widely dispersed in the Americas as corn.

Kiwhicha was the caudatus strain that adapted itself to high altitudes, desert soils and truly challenging conditions.

The plant grows vigorously and tolerates drought, heat and pests. It adapts quickly to new environments, many of them inhospitable to conventional grain crops.

The small grains of kiwhicha are produced generously by long colorful seed heads, sometimes over 100,000 grains per plant. Like all amaranth seeds they are flavorful and nutritious.

Kiwhicha can be popped, malted, baked, puffed into breakfast food, boiled into porridge or ground into flour. Kiwhicha is one of the most sustaining grain crops with more protein than most major cereals and an amino acid balance close to nutritional perfection with high levels of lysine (usually deficient in plant protein), calcium, phosphorus, iron, potassium, zinc, Vitamin B complex and Vitamin E. It is considered an excellent food for invalids, children and the elderly.

Local Importance: Amaranth species, including Kiwhicha, are undergoing a renaissance in their native range. Amaranth varieties have become specialty crops in the United States especially in areas that have had to return to dry land farming. The grain is esteemed as health food and breakfast food.

Global Importance: Kiwhicha is crop candidate for arid, saline, unstable areas and harsh climates. Kiwhicha and other amaranth varieties could tip the scales in the battle against famine in the dry marginal lands of India, China, and Africa.

Give us this Day our Daily Bread This ancient prayer is the truest expression of the hungry. The grains mentioned in this chapter are an answer to that prayer. We cannot neglect them if we are to achieve abundance.

Modern man has domesticated few food crops; some scientists would say that modern man has domesticated **no** new food crops. What we eat today we owe to the efforts of our ancestors, efforts which began in Neolithic times.

To quote *Shattering: Food Politics and the Loss of Genetic Diversity*:

> *For all our technological wizardry, we human beings still owe our existence to a few inches of topsoil, an occasional thunderstorm and a handful of crops. (Fowler and Mooney, 1990, pg. 3)*

Approximately 300 food plants have undergone the entire process of domestication and only 130 actually feed the world's population. Modern agriculture recognizes but a dozen of these as "big business" and within those few only the varieties which fit into a high input cropping system.

Needless to say a system of this kind is not prone to the sort of experimentation that will allow the domestication process to continue. This only happens on the fringes of agricultural endeavor and is usually done by the indigenous people who live in marginal areas.

There is no better way to invigorate an increasingly static and impoverished genetic base than the introduction of diverse, wild, unmapped genes. This may be the perfect time in human history for a Second Domestication in which we add another hundred crops to our agricultural treasure chest.

Simple technologies have eliminated the problems posed by small grains and grains which need to be husked or decorticated. Refrigeration and better storage facilities allow more perishable food to be kept. Advances in milling, cooking, baking and fermentation have totally changed what we can do with raw foods when we get them.

Why do we settle for a fraction of the planetary crop potential? There are hundreds of wonderful wild plants that we can tame and eat or manage in their indigenous stands to move us closer to food security.

Some Misconceptions about Wild Plants

1. If they were "any good" someone would be growing them by now. Plants that are utilized in the wild state, though they may be splendid food, are usually taken for granted by the "natives" and overlooked by the scientific establishment. They may have a very small area of natural dispersion and so are known only to the few inhabitants of the plant's range. In addition to this until the plant has been proven to have some kind of commercial value there will be no attempt to domesticate or investigate the plant. This is the reason why many wonderful foods have remained local wild crops.

2. The crops people were "meant to eat" have been with us since time immemorial. What should be remembered about our modern diet is that many of the plants we depend on were unknown in the Old World as little as four hundred years ago. These include corn, beans, squash, peppers and potatoes. Other foods such as oranges, bananas and avocados only became available to the populations of the developed world in the 20th century. Soybeans became an important crop after WWII. Kiwis, pitayas, kiwanos, passion fruits, sunchokes, arugala, bok choi and most oriental vegetables have only come to the marketplace in the last twenty years. There is no reason why this expansion of choice should only apply to fruits and vegetables. New tuber and grain crops could do much to correct the deficient western diet and create opportunities for people who would like to return to agriculture. New herbs, spices, beverages and condiments could make the staple diet of the poor more wholesome and appealing.

3. Domesticating a plant takes decades, if not hundreds of years. Actually domesticating a plant, that is finding strains of a plant that produce well under cultivation and figuring out the best way to cultivate it, can be accomplished very swiftly if time, care and some funds are invested. This process usually starts with a survey of the plants in the wild and a collecting expedition to acquire interesting genotypes. These genotypes are ordinarily planted out in various experimental formats. Then the types which respond favorably to cultivation, produce foodstuffs of good quality and are well adapted to the target area are selected. These promising genotypes are the raw material for conventional plant breeding which then progresses according to the life cycle of the plant.

4. When we have things like wheat and apples, wild plants are not "worth the effort." It has become glaringly obvious that the whole planet cannot eat wheat and apples. Even if it were possible it would not be a good idea. Overuse of the same few foodstuffs is not good for human health as the climbing rates of food sensitivities and allergies have shown. There is a considerably body of evidence to suggest that different races and ethnic groups should not be consuming the same types of foods, especially not the typical western diet which does not appear to be good for anyone. (Itzkovich, 2000) Examples of this new kind of thinking can be found in the revolutionary diet programs for Native Americans of certain Southwestern tribes and for the First People of Hawaii.

For many of them the western diet causes high blood pressure, diabetes and obesity. A return to diets based on roots and tubers, whole grains, vegetables, fruit (and in the case of the Hawaiians, fish) has restored many obese, sickly individuals to health. They were much more obviously unwell than the descendants of the Europeans probably due to hereditary differences in the way their bodies store and process sugars and fats.

Why Now?

Many potentially useful plants have never been domesticated. This does not mean that they are not used as food. Sometimes these plants are gathered from the wilderness or from a semi-wild common area. Often the seasonal bounty of a stand of plants will be picked, eaten or turned into a preserved product like jelly or wine.

Cloudberries, oregon grapes and blackberries are common examples of useful plants that are utilized seasonally in this manner.

While most plants of this type have the potential to be minor local crops at best there are some wild plants that produce food of a quality and quantity that demands attention.

A few of these wild wonders are being domesticated now, a process which sheds light on our agricultural history.

These wild plants include grains, palms, melons, pods and berries. All produce better food than many similar domesticated plants with much less water, few inputs and little or no attention on the part of mankind.

Why domesticate them now? **Because we need them.**

Channel Millet: Food from the End of the World

This little known wild Australian plant Channel millet (*Echinochloa turnerana*) grows exclusively in the channel country of inland Australia; a desolate eroded area cut by the sun and wind into gullies called "channels." Vegetation is rather sparse, but in some channels stands of this wild cereal manage to stay green well into the dry season.

About 100mm of rain falls annually in one or two downpours filling the channels with water and soaking the ground. When the water dries out the mud becomes a cement-like crust.

Channel millet always grows in silty deep clay that has been saturated with water. Deep flooding is necessary for germination but one good soak is enough to allow the millet seeds to sprout, grow and form a mass of dense grass with heavy seed-heads. There is a strong resemblance to conventional millet in the stalks and heads of channel millet. Like millet the wild grain is much sought after by small birds, which pick the grains out of the seed heads while perched on the stalks. The grain is also eaten by sheep, cattle and horses. The whole plant makes good hay.

Current Use: Sometime fodder and browse in channel country.

Potential Uses: Fodder, browse, arid lands grain and hay crop.

Research Needs: Seed collection, basic botanical studies, testing with seeding, harvesting and grain cleaning equipment, exploration of bread making potential.

Pejibaye: The Peach Palm

The pejibaye palm (*Guilielma gasipaes*) is little known outside of Central America. But the pejibaye fruit has twice the food value of the banana and can provide more carbohydrates and protein per acre than corn. The spiny multiple trunks of this extraordinary palm emerge from a common root. The two to six cm fruit is borne in large date-like bunches weighing between ten and 12 kilos.

There may be a dozen bunches on a tree and the tree fruits twice a year when there is adequate moisture.

In short it is possible to harvest over 200 kilos of fruit from each tree annually. This is a harvest index approached by no plant of the palm family except for the date palm.

The fruit is not sweet like the date. It is starchy and chestnut like. The fruit is usually boiled in salted water. It can be eaten after it is cooked or dried for later use. The fruit can also be roasted like a potato and contains generous amounts of carbohydrates, protein, minerals and vitamins.

The white hard-shelled seed is also eaten or pressed into oil. This part of the fruit is coconut like and can be dried.

The palm begins to bear fruit when it is between six and eight years old and can continue to fruit for 80 years. It is well adapted to heavy clay soils, preferring areas where the annual rainfall is 2,500 mm or less.

Current Use: Local crop in Costa Rica and the lowlands of South America.

Potential Use: Major food crop for the humid tropics.

Research Needs: Search for superior trees, propagation experiments, experimental plantings, cooking and processing experiments for the fruit and oil seed.

Topaz Fruit: the Tastiest Cactus Fruit

This strange climbing cactus (*Selenicereus megalanthus*) is a relative of the "strawberry pear" of Central America (colorplate 18 on page 267). It is also called the yellow pitaya.

The plant itself is a spiny, viney, fleshy creeper, light green in color which climbs up rocks, trees or trellis and supports itself with fibrous aerial roots.

Large buds on long pinkish stalks soon burst into orchid like cream and white flowers. Like most cactus flowers they are short lived. If pollinated a fruit forms, bright yellow and thorny on the outside. The thorns loosen and can be brushed off when the fruit is ripe.

The ripe fruit, usually between 200-300 grams, is filled with clear pulp studded with flat brown seeds. It is fragrant, refreshing and delicious, probably the tastiest of all cactus fruits.

Current Use: Minor local crop in Central America.

Potential Use: Exotic export crop that needs little water.

Research Needs: Identification of superior and disease resistant plants, experimental plantings, Integrated Pest Management (IPM) and pest control research.

Buffalo Gourd: Seedgourd of the Badlands

The buffalo gourd (*Cucurbita foetidissima*) is a sprawling vigorous perennial native to the deserts of Mexico and southwestern US. The plant needs warm dry weather and dry soil with light seasonal rains to thrive. It produces an abundant crop of seeds rich in proteins and oil. Its huge forked tubers burrow deeply into the ground to obtain and store water. The seeds are surrounded by juicy pulp in hard yellow round fruits. The fruit can be harvested by combine. In arid areas the fruit becomes dry and brittle and the seeds can be threshed out.

Each fruit contains between ten and 14 grams of seed. The seed contains between 30 to 35 percent protein and over 30 percent oil. Plants produce between 50 and 80 fruits per season. The seeds can be crushed to produce polyunsaturated oil for food and industrial use. The huge tubers are recognized as a valuable source of starch.

The estimated yields of the buffalo gourds compare with those of peanuts and other oil-bearing crops yet the buffalo gourd has not been improved by modern breeding.

A few years of research and development might bring this interesting plant out of the wilderness and into cultivation.

Current Use: Native American "lost crop." Researchers began to investigate this plant in the 1970s.

Potential Use: Oil, starch and protein from arid areas.

Research Needs: Identifying superior plants and propagating them, basic botanical research, basic nutritional research, long-term toxicity experiments.

Zarzamora: Quite a Mouthful

The zarzamora or "wild bush berry" is also called the Columbian Giant Blackberry (*Rubus macrocarpus*). This berry is too big to fit in a person's mouth, sometimes as large as a hen's egg. Native to a narrow, inaccessible upland zone of Columbia, the canes and flowers of the plant resemble those of an ordinary blackberry. The berry itself resembles a huge dark raspberry usually five cm long and two to three wide with a sweet loganberry like flavor, wine red color and slightly acid pulp. This interesting Rubus species is being domesticated now, a process which has led to the establishment of plantations, experiments in hybridization and attempts to establish the zarzamora outside of its native range. So far the experiments to grow this species outside its natural area have failed. Attempts to hybridize the zarzamora with cultivated Rubus species have not yet been successful.

Interestingly enough there is another giant berry in South America, the Mora de Castilla (*Rubus glaucus*). This plant is native to a large area stretching from the northern Andes to the southern highlands of Mexico. In Columbia it has become a cash crop. It is superior in taste, size and keeping qualities to most commercial blackberries and raspberries. The plant is a vigorous, lush shrub that produces fruit year round in moderate climates. Twenty tons per hectare have been reported from plantations in Columbia making the Mora de Castilla an important local crop. But the berry is almost unknown outside of the described area.

Current Use: Local specialty crop.

Potential Use: Exotic fruit crop with good export potential.

Research Needs: Is the zarzamora sensitive to photoperiods? Genotype collection. Hybridization between the zarzamora and easier to cultivate Rubus species (*Rubus roseus* and *adennotrichus* for example) from the same climatic zone.

Andean Blueberry: Sweetness from the Paramos

This blueberry (*Vaccinium meridionale*) is almost black in color, round, about one cm in diameter and borne in clusters of between ten and 15 berries on a small tough tree with reddish bark and dark green leaves. This tree grows in the windy highlands of Columbia called the Paramos usually appearing at 2,000 meters and able to survive to about 4,500 meters. It also grows in the mountains of Jamaica at 1,000 to 2,000 meters in altitude. The Jamaican trees are generally believed to have been an 18th century introduction but there is no clear record of who introduced the plants. The skin of the fruits is rather tough and the Andean blueberry seems to have a longer shelf life as well as being larger than the conventional blueberry. The fruits are popular in preserves, pastries, wines, ice cream, cakes and frozen desserts.

Current Use: Local wild crop.

Potential Use: Berry crop, added value crop for chilly areas.

Research Needs: Seed collection, propagation experiments, genotype selection, experimental plantings.

Lalob Trees: The Spiny Treasure of Desolation

The lalob or desert date (*Balanites aegyptica*) is a tree with thorny drooping branches and a distinctive white and green trunk. It is common in the deserts of Africa and the Middle East where it grows in sandy soils and thrives on hidden ground water. It is also a tree of the wadis and the *sabha* or saline marsh thriving where seasonal floodwaters gather and sink into the ground.

The fruit of the lalob resembles a date with a large pit, gummier in texture but full of sugar. The kernel inside the seed is edible and rich in oil and protein. Cooking the fruit makes it more palatable and the fruit paste is sometimes made into jam. The roasted kernel is very tasty as is the oil which can be pressed from it. The flowers of the lalob are edible boiled or fresh. The young leaves can be boiled and eaten like spinach. Extracts of the bark are used for purifying water as they do not harm people or animals but are deadly to a number of water borne parasites including the ones which cause schistomasis and dracunculiasis.

The lalob has medicinal value and has been esteemed as a cosmetic, antiseptic and anti-inflammatory since Biblical times. According to the Bible the Patriarch Jacob sent lalob sap and other rare substances as a present to Pharaoh when he asked permission to buy food in Goshen. Lalob sap is one of the ingredients of the famous balm of Gilead, one of the tree's many nicknames is the Jericho Balm tree.

Lalob sap was also valued as an aromatic and used in incense. The globs of fresh resin resemble sweets and are as pleasant to the taste. Sap and fruit can be fermented together to made a bittersweet liqueur. The seeds are also rich in the raw materials for the production of steroids.

Lalob is useful as a forage and fodder plant since animals eat not only the leaves but also the seeds and fallen fruit. The tree provides shelter, shade, living fences and islands of green in some of the most dry and desolate areas.

Despite all these important and useful properties the lalob is still a wild tree, underexploited as a source of food and fodder, ignored by all but a few herders and desert farmers.

Current Uses: Local wild crop, local forage plant, local medicinal plant

Potential Use: Fruit and processed fruit products, edible oil, cosmetic oil, raw materials for steroids, liquors, water purifying substances, nuts, leafy vegetable, fodder plant, anti-erosive and anti-desertification, living fence, building material, windbreak and shade plant.

Research Needs: Selection of best wild types, continuing investigation into uses and properties of the plant, experimental plantings.

Nipa Palm: Sugar Palms of the South Pacific

The Nipa palm (*Nypa fruticans*) grows in the tidal marshes and on submerged banks of bays and estuaries in much of the South Pacific. Wild stands are found in West Bengal, Burma, Malaysia, Australia, the Philippines, New Guinea and Indonesia.

There are some cultivated plantations but since rice can be grown on the same type of land most Nipa sugar is gathered from wild trees. The sap is collected from the mature fruit stalk which is cut and tapped. If this is done carefully it is possible to collect between 30 and 50 liters of sap per tree per season. Cultivated trees reportedly produce approximately half a liter of sap per day for ninety days.

The sap begins to ferment immediately when tapped. It is almost 15 percent sugar and the presence of wild yeasts will lead to the end product of acetic acid. The sap can be used to produce ethanol or boiled into an edible syrup. Working in the swampy terrain of the wild stands is sometimes done by boat with the sap collectors cutting the palm branches selectively to open up passages by which they can approach the trees. The fronds and fiber of the Nipa palm are used as thatch and cordage.

Current Use: Interesting underutilized sugar and energy crop from tidal wetlands.

Potential Use: Sugar and energy crop for swampy areas and tidal flats.

Research Needs: Identification of superior trees, propagation experiments, genotype collection, more efficient sap collection, preservation of wild stands.

Mastic: The Gum Trees of Chios

The product of the mastic trees (*Schinos pistacia*), chios mastic, is not precisely a food though it is used in food preparation. It is used in ink-making, printing, medicines, electronics, baking and art supply manufacture as well. It is gathered from the wild green valleys of Chios, an island blessed with a wetter and more moderate climate than most Greek isles, where the mastic shrubs grow on terraces among the volcanic rocks under startlingly blue skies.

The trees are small with dark green shiny, red-veined leaves. They are carefully pruned to allow access to the twisted trunks. White sand is spread around the base of the tree and then the tree is marked with a series of deep, tapering diagonal cuts. The strongly scented gum drips down the trunk onto the sand or dries in long strands and beads on the dark bark. The smell around the trees is strong, almost turpentine-like. The sap itself is sticky when fresh and hardens into a whitish mass within a few days. Gum collected from the sand is washed and rolled into balls.

Good quality mastic is worth sixty dollars per kilo. A small box can be purchased for a dollar's worth of local money and chewed like chewing gum. It resembles chewing gum in texture but has a strong piney taste.

Current Use: Unique local wild crop. The trees are found in small groves and rows along ancient terraces in the south of the island, but local sap gatherers swear they didn't plant them there. The trees "plant themselves."

Potential Use: Cash crop for areas with adequate water but too rough for cultivation. Possible alternative to olives, figs, cork and other Mediterranean crops.

Research Needs: Find out why plantings of Chioan mastic trees have failed on other Greek islands.

Naranjilla: Culinary Delight of the Andes

The naranjilla (*Solanum quitoense*) is a relative of the tomato. It is extremely versatile and can be eaten raw, cooked, used to make juice, jams, jellies, pies, cakes and confections. In Central America it is sometimes added to yogurt and ice cream. The fresh juice is turned into frozen concentrate and can be made into wine.

The name naranjilla means "little orange" in Spanish but the plants and fruit are not citrus like despite the similarity in color.

The naranjilla has long been used as food but it is only recently that the crop can be considered a domesticated plant. There is no evidence that it was cultivated before the Spanish conquest so the domestication process must have started in the last few hundred years. The process of breeding and refinement is still in progress in the plant's native range. What happens there will determine if the amazing taste of the naranjilla will become known to millions or whether the naranjilla will remain an obscure minor crop.

The fruits are picked when they begin to show color and are covered with an irritating fuzz which can be wiped off when the fruit ripens. The fruit has a shelf life of about two weeks without refrigeration. The yellow skin and greenish flesh are acid and tangy sweet. As a source of fruit juice the naranjilla is potentially as popular as citrus fruits.

The leaves of the plant are large and serrated, sometimes speckled, sometimes with a purplish bloom. The plants mature swiftly and begin to bear fruit at the axils in about a year. Propagation by stem-cutting or tissue culture is not difficult and seeds germinate easily. Plants bear fruit for several years if planted in well-drained soils in areas of moderate climate.

With all these attractive qualities it is hard to believe that the naranjilla has remained an obscure minor crop. Formidable pest problems in naranjilla monocultures have kept the naranjilla out of mainstream agriculture. The naranjillas are vulnerable to root damage from nematodes, bacteria and fungal infections. The branches of the plant are brittle and easily damaged by the wind. Because of this the plants are often intercropped with the tamarillo, the banana or the achira. Experimental planting between babaco, mandarin oranges and passionfruit have been successful so perhaps the naranjilla does best as an intercrop with heartier neighbors as nurse trees.

There is little genetic diversity in the cultivated strains of the plant, which indicates that an examination of wild relatives may be advisable. There are wild naranjilla types from Peru to southern Mexico and a cold weather race which is found at higher altitudes in Columbia and Venezuela.

The climatic requirements of the cultivated naranjilla are narrow. The area in which they are grown must be frost free but not too hot. Naranjillas require a short daylength.

Bringing this unusual and useful fruit into mainstream agriculture is not going to be easy. But the unique flavor of the naranjilla and its versatility make the effort worthwhile.

Current Use: Local minor crop.

Potential Use: Major juice crop, flavoring, base for value-added products and prepared foods.

Research Needs: Hybridization with wild relatives, disease resistance, cold tolerance, investigation into daylength sensitivities, continued intercrop experiments.

Kram-kram: The Terrific Tumbleweeds of the Sahel

On the border between the Sahel and the Sahara there are massive stands of a dune stabilizing wild plant called kram-kram (*Cenchrus biflourus*). As might be expected of a wild plant in such a challenging area kram-kram is very tough, drought tolerant and salt tolerant. In its juvenile state is a valuable forage plant. Later as the plant matures animals avoid it because of its spines. The grains are enclosed in a truly vicious sand burr with many sharp hooked thorns. Burrs cling to clothing and the fur of animals. When the burrs mature in large quantities they begin to roll with the wind, clinging together in spiky masses.

Indigenous inhabitants collect these tumbleweeds with giant combs or brooms, pound the entire mass in a mortar and free the tasty white seeds from the spines. The seeds contain nine percent fat and 21 percent protein, making kram-kram a true powerhouse of food energy. The seeds are then crushed to make porridge or parched and stored for later use.

There are spineless types of kram-kram and types which contain an impressive 25 percent protein, more than double the amount of protein in corn and wheat.

Kram-kram could be domesticated in a few years, giving the inhabitants of the Sahel food for themselves and for their animals.

Current Use: Wild crop, browse, famine food.

Potential Use: Major food crop for arid and saline lands, water-saving food and forage crop.

Research Needs: Crossbreeding with spineless varieties, investigation of bread-making potential, identification of superior strains, seed collection.

Narra Melon: The Forbidding Fruit of Namibia

The narra melon (*Acanthosicyos horrida*) grows on deep sand dunes in the Namib desert which spans parts of Southwest Africa, Namibia and southern Angola. It is a spiny leafless shrub with a an equally fierce-looking fruit. Narra fruits can be as large as melons or as small as billiard balls. The young fruit is bitter but when ripe the narra is sweet and palatable with a cream colored or light yellow flesh and a fresh pineapple like taste.

The large white seeds of the narra are delicious and sold as "butterpits" in South Africa after having been roasted and salted. The seeds are also used in baking and as a substitute for rennet. The oil from the seeds may be valuable in cosmetics or cooking as it is flavorful and rich in protein.

The narra melon is drought tolerant and can survive years without rain. Though the fruit is used by indigenous people and avidly eaten by animals little has been done to advance the domestication of the narra.

Current Use: The fruit is eaten by local Bushmen and animals. The seeds have become a local snack food.

Potential Use: New fruit crop for arid and saline area, possible source of edible oil with unusual characteristics.

Research Needs: Seed collection, cultivation trials, identification of superior types.

The Strengths of Wildness

These plants survive and produce food in challenging areas unsuited for conventional agriculture. Many are resistant to greater stresses than domesticated plants could bear. Some have unique features of adaptation. Some carry valuable traits that may allow them to become as useful to humankind as wheat or apples.

With our most important crops losing the strength of diversity we need the newness and the wildness of these plants to help us achieve abundance.

VIII. The Fruits of Bitter Waters

The sea has been an important source of food since time began for humankind. Fish, shellfish, crabs and mollusks have been harvested from the waters that cover two thirds of the globe. Mankind has hunted seals, walruses, otters, dolphins and whales, some of them to the point of extinction.

In the underwater plant kingdom mankind's depredations have not been so severe. Several types of seaweed, kelp and algae are esteemed as food, especially in the Orient. Other types of sea plants and kelps have been used as fodder, fertilizer and bedding for animals.

There is a third source of food from the sea that has recently begun to attract attention. These are the rare plants that are rooted firmly in the earth but can be cultivated and irrigated with seawater.

These salt loving plants are capable of producing fruit, fodder, essential oil, tubers, fresh vegetables, wood and grain.

Three possible domains for salt tolerant crops might be farmlands salinized by poor irrigation, areas that lie above reservoirs of brackish water and coastal deserts.

In a hungry world the use of such plants could save precious sweet water for drinking. Salt and tidal flats could be cultivated. These plants could supply valuable genetic material which could be transferred to salt sensitive food plants to make them hardier.

Plants That Love Salt

Halophytes are plants that grow in soils or waters which contain significant amounts of salts.

Halophytes have strategies for dealing with excess minerals. The tamarisk tree absorbs salt and excretes it in falling leaves. These salty needles pile up around the base of the tree effectively choking out the competition.

Some halophytes allow the salt to invade them, becoming saline themselves in the process.

Some halophytes possess stress genes, which allow them to survive concentrations of salt close to that of seawater.

Other halophytes can change the pH of the saline water by exuding natural acids.

It might be possible to transfer some of these traits to cultivated populations of plants therefore making them tolerant of salt.

Acacia: The Ubiquitous Tree of the Desert

Acacias grow in the Australian outback (colorplate 19 on page 267). They can also be found in the deserts of Africa and the barren reaches of the Sinai desert as well as the arid areas of Central Asia. What is not generally known is that many acacias have edible seeds, leaves, bark and sap. Some seeds are rich in nutrients with higher energy, protein and fat contents than wheat or rice. The high protein contents suggest that some acacia seeds have bread making potential. The high fat content hints at the acacia's potential as oilseeds. In any case this super hardy tree which thrives where no other vegetation can survive could be the ancestor of new lines of food plants.

Table 3: Acacia Seed Composition

A. anuera	2220 kJ of energy, 23.3% protein, 37.0 fat, 25.5 carbohydrate, 9.7 ash
A. corica	1491 kJ of energy, 23.8% protein, 7.7 fat, 48.1 carbohydrate, 3.7 ash
A. cowleana	1507 kJ of energy, 22.2 protein, 10.1 fat, 44.6 carbohydrate, 7,2 ash
A. dictyophleba	1519 kJ of energy, 26.8 protein 1.9 fat 79.9 carbohydrates, 6.5 ash

Source: Peterson, 1978

Other Acacia Species Used As Food

Acacia farnesiana

This species is known as the cassie flower. It is a shrub up to four and a half meters high indigenous to tropical America but naturalized in India and tropical Asia. The gum of the cassie flower tree is suitable for confectionary and the leaves replace the tamarind in chutney making. The flowers are used in brewing a medicinal tea.

Acacia leucophloea

This species is native to the dry hills of the sub-Himalayan areas.

It is a moderate sized spiney deciduous tree with shaggy bark and flat shiny brown pods. The gum is used after the manner of gum arabic. The seeds are edible and the bark is sometimes pounded and eaten in times of scarcity as well as being used to prepare spirits of palm from sugar and palm juice.

Acacia nilotica

This species is a moderate sized tree with fissured bark, delicate pinnate leaves and distinctive yellow puffballs of flowers. It is indigenous to the North Deccan area of the Indian subcontinent but has spread to the rest of tropical Asia and some countries in the Middle East. This species is believed to be the first source of commercial gum arabic though now A. senegal is the preferred source in India. In times of scarcity both the seeds and bark are ground and eaten.

Acacia senegal

This species is a three to six meter tree with grayish green bark and thin pods containing five to six seeds. This species of acacia is the preferred species for the production of the odorless demulcent and emulsifying agent known as "gum arabic." The tapping is done by making incisions with an ax and collecting the gum when the drops of sap harden. The trees of the species Acacia catechu are tapped in a similar manner. In times of scarcity the green pods of A. senegal may be eaten as a vegetable.

Acacia sinuata

This species is a climbing shrub with thorns and feathery leaves which occurs in jungles and the hotter parts of India. It flowers generously and the flowers are small and form under the leaves in round masses. The tender acidic leaves are used in chutneys and the bean like seeds are roasted and eaten.

Local Importance: Wild crops in the most desolate areas of the earth.

Potential Importance: New source of protein, tertiary food and oil for extreme environments.

Salvadora: Fruit, Soap and Forage from Desolate Shores

Two species of salvadora produce edible olive like fruits and seeds which can be pressed into high quality oil for soap making. Both trees are evergreen, densely crowned and shrub like, sometimes the only green plant in an otherwise barren beachscape. Both salvadora species are cultivated trees in India and Pakistan where they provide wood and fodder as well as fruit and oil. Salvadora persica grows on Indian coasts just above the water line. There is some evidence to suggest that the seeds are spread by ocean currents. The intensely green shrubs thrive just out of the reach of the tide.

Salvadora oleoides is grown on some of the worst salt affected lands in Pakistan, though the trees are usually replaced by cash crops with the introduction of channel irrigation. The fruits can be consumed without leaching as they are far less bitter than the olive, they are yellow when ripe and contain glucose, fructose and sucrose. Timber from both trees is hard and water resistant esteemed in local woodcarving and furniture industries.

Local Importance: Utility crop for desolate coasts and areas too saline for other food crops.

Potential Importance: Oil, forage and fruit for subsistence and sustainable projects in saline and coastal areas. Land reclamation tree for salinated areas.

Seagrapes: Sweetness from Salt

The sea grape (*Coccoloba uvifera*) is a dwarfed and bushy tree that forms colonies on windy, spray swept beaches. In this situation the sea grape rarely reaches more than two and a half meters in height. In sheltered, less saline areas the plant can become a spreading solitary tree five meters tall. In both cases the plant produces quantities of grapelike juicy fruits which can be eaten fresh or turned into new wine. On the islands of the Caribbean, the sea grapes native range, the trees also provide durable wood for furniture and cabinet making. The abundant flowers of the sea grape supply high quality nectar for apiaries. The fruits are sold in the marketplace as a local wild crop. Their thin skins and sprightly taste make them a seasonal treat. They are also tasty in salads and used in making soft drinks.

Local Importance: Local specialty, beverage plant, bee culture, high quality wood.

Potential Importance: Possible crop for sandy shores, windbreak, anti-erosive.

Indian Almonds: A Nut Tree Spread by the Ocean Currents

The Indian almond *(Terminalia catappa)* originated in Malaysia and was spread by fruits borne on the ocean currents to many shores and islands. It has adapted so well to India and Burma that it has become a cultivated tree. Wild plants have become common in Africa, the Pacific islands and the coast of tropical America.

The ellipsoidal fruit with its edible kernel is usually washed up on a beach above the tide level. There it takes root and grows into a sturdy hardwood tree which can withstand storms, salt spray and flooding. The kernels are very much like the almond. There are sweet palatable varieties and other types which are too bitter to be edible. The nuts are high in protein and sometimes pressed into edible oil.

The tree is particularly suited to sandy and rocky coasts. Its bitter foliage is not high-quality fodder but sheep and goats will browse on new growth.

Local Importance: Nut crop in India and Burma. Source of hard, durable wood.

Potential Importance: Possible nut crop for sandy, rocky shores and salt flats in warm areas.

Tamarugo: The Unkillable Legume

The tamarugo (*Prosopis tamarugo*) is native to one of the most desolate areas in the world, the Atacama desert of northern Chile which is notorious for rapid changes in temperature and covered thickly with crusts of salt several feet deep. Droughts can last up to seven years in the Atacama and the desert is swept by alternately searing and chilling winds. Few plants can survive in this wilderness but the tamarugo not only thrives but produces nutritious pods, seeds and leaves that can be eaten by animals. Tamarugo pods should be eaten by humans only after thorough leaching and then resemble the carob in taste and texture.

Tamarugo trees have been planted in Chile to replace the vanishing native tamarugo woodlands. This is done by cutting holes in the salt crust and planting the seedlings below the surface. This underground planting method allows the quickly developing tap root to reach the brackish water table and protects the young plant from the scouring winds. Pilot plantings have been introduced to similarly inhospitable areas with extremely saline soil, high brackish water tables and damaging winds. Tamarugo plantations can support goats and sheep as well as most high quality pasture land.

Local Importance: Recognized as a botanical treasure and replanted in the Atacama desert of Chile to produced forage and fodder.

Potential Importance: Possible source of forage and shelter for animals, windbreak, land reclamation, anti-erosive for saline lands.

Salicornia: When Seawater is Less Salty than the Soil

Salicornia species (*Salicornia sp.)* are cultivated in several countries on salt flats where seawater has been used to reduce the level of salinity. Needless to say this indicates that some salicornias are **very** salt tolerant.

Salicornia bushes produce an abundance of small seeds which can be pressed to produce a safflower-like oil. The rest of the plant and the oilcake residue can be used as animal feed. There has also been some success in producing leaf protein from selected species.

Plantings in Mexico produce oilseed at a yield of two tons per hectare and fodder at a yield of close to 20 tons per hectare with sea-water irrigation.

Salicornia has been planted in Egypt and Saudi Arabia in especially saline areas.

Local Importance: New oil and fodder crop for otherwise unusable land.

Potential Importance: New source of oil, fodder, leaf protein for very saline areas. Land reclamation plant.

Zostera marina: Wheat in a Raincoat

Eelgrass *(Zostera marina)* is one of the only known plants that roots, grows, flowers and seeds completely submerged in salty water. It grows well in shallow, sunny waters, forming a dense grassy carpet on the ocean floor and binding the silt and sand.

The plant is adaptable and can be found from the sub arctic to the temperate coasts of the world. There are related species native to the southern hemisphere.

Growing grain in the sea is a new idea but the zostera species' vigor and diversity may make this possible.

The wheat like grain, usually three and a half mm long and one and a half mm around, can be threshed out of the seed head by pounding and shaking dried plants. This causes the seeds to drop out of its husk, which has been likened to a "raincoat." After winnowing and separation the grain was used by Seri Indians to make unleavened bread, gruel and honey cakes.

The somewhat bland flavor is reminiscent of wheat and the protein content approximately 13 percent compares favorably with most modern cereal crops. The grain also contains approximately 50 percent starch and one percent fat.

Local Importance: Lost crop of the Seri Indians.

Potential Importance: New grain crop for shallow estuaries all over the world, shore protector, shore stabilizer.

Palmer Salt Grass: Bread that Comes in with the Tide

Palmer salt grass (*Distichlis palmeri*) grows in tidal flats and thrives on inundations of salt water. Unlike the Zostera species which must remain submerged, the tidal flats are accessible for part of the day allowing for some rudimentary cultivation. It is a sturdy perennial plant spreading by tough rhizomes which send up spiky shoots about half a meter tall. It is fast growing and once planted does not need to be replanted.

Disturbance of the tidal flats by storm surges or high tides often serve to spread the grass rather than curtail it as the rhizomes take root wherever they drift. The spikelets which bear the seeds of the plant dislodge when ripe, float and are carried into shore with the tide.

Yields of a ton per hectare have been reported from natural stands, but more could probably be collected if efficient collection methods could be developed.

The seeds can be made into bread.

Local Importance: Wild grain crop of a specific locale.

Potential Importance: Minor cereal crop for tidal areas, shore protector and stabilizer. Land reclamation plant.

Bajra: Millet Watered Once

Pearl millet (*Pennisetum glaucum*) is a popular source of food in India and Africa. Extending its range of cultivation will make more food available for the poorest of the world's people.

Millet is usually grown in areas with adequate rainfall to allow for seasonal cropping. In some areas it is planted on saline land and irrigated just enough to allow the plants to develop.

What is not generally known is that bajra millet seedlings can be established with fresh water then irrigated with seawater until they reach maturity.

Crops irrigated in this manner can yield between one and two thirds and two tons of edible grain per hectare, making bajra an important source of food while saving much fresh water.

Local Importance: Cereal, fodder for seaside areas in India and Africa.

Potential Importance: Major grain crop for seaside areas in the developing world, fodder, shore protector.

Mallow: A Little-Known Edible Seed

Seashore mallow (*Kosteletzkya virginica*) can be sprouted in fresh water like bajra millet then irrigated with saline water. It is a perennial plant which survives five years in cultivation. The plant can tolerate two to three percent salinity.

Hulled seeds resemble millet and contain as much as 32 percent protein and 22 percent oil making these little known seeds superior to most modern food crops in their energy potential. Grain yields from experimental plots irrigated with water containing two and a half percent salt have yielded up to one and a half tons per hectare.

Local Importance: Wild fodder and seed crop.

Potential Importance: Unknown

Bush Coral: A Stranger Just Above the Tides

Bush coral *(Tecticornia sp.)* are native to Australia where they grow in coastal mudflats above the tidal level. While the branched, warty looking plants look more like exiled sea plants than a potential cereal crop they produce small edible seeds that the Australian aboriginal peoples used to make unleavened bread. While it is not known if the aboriginal people actually planted tecticornia plants they gathered the seeds regularly, returning to the coastal flats before the seasonal rains to utilize this source of food.

New plants grow from seeds when the areas are leached by the rain, then survive by absorbing salty water from the wet earth of the mudflats.

Local Importance: Lost wild crop of very marginal areas.

Potential Importance: Unknown. Should be investigated.

Wheatgrass and its Relatives: Propping up the Staff of Life

Wheat is one of the most important sources of human nutrition. Anything which may extend the range, improve the nutritional quality or increase the hardiness of the wheat plant is worthy of serious attention.

Many wild wheat relatives show an astonishing amount of salt tolerance and can be irrigated with water straight from the sea.

Two species, tall wheatgrass (*Elytrigia elongatum*) and jointed wheatgrass (*Elytrigia pontica)* can survive concentrations of salt greater than that of seawater. Salt tolerance may be transferred to conventional wheat from such wild relatives.

Couch grass (*Thinopyrum bessarabicum*) has been crossed with wheat to produce a fertile hybrid that is salt tolerant and sets seed.

Local Importance: Few wheatgrass species are used or cultivated.

Potential Importance: The genetic material from these plants may revitalize and invigorate wheat varieties all over the world.

Wild Water Chestnuts: Gourmet Food from Saline Swamps and Marshes

Wild water chestnuts (*Eleocharis dulcis)* are smaller and harder than the cultivated types. They grow in salt swamps and salt mashes with little attention from the local people who gather them. They are valued as delicacies and can be added to soups, stews and salads or pounded into meal.

They are found in salty wetlands in Asia and show wide diversity in color, size and flavor. They may be useful as a new minor tuber crop for wet, saline areas or their salt resistance and hardiness could be transferred to cultivated varieties.

Local Importance: Local specialty, wild crop.

Potential Importance: Minor tuber crop for wet, warm saline areas. Valuable genetic material to invigorate cultivated water chestnuts.

Saltworts: Seventeen Tons of Roots per Hectare

Seventeen tons of saltwort (*Batis maritima)* roots per hectare have been produced by using seawater to irrigate this odd desert plant.

Saltwort roots were used by the Seri Indians as food. They baked the tuber, boiled it or ground it for its edible starch. Saltwort roots are said to have beneficial effects on the diabetes and obesity problems of southwestern Native Americans.

Many saltwort tubers are fibrous but others have the crispness of potatoes. Selection and cultivation of this plant could result in a new major tuber crop for areas that are arid, saline and hot.

Local Importance: Rediscovered lost wild crop.

Potential Importance: Possible new tuber crop for deserts, seaside areas.

Sea Fennel: Salad from the Sand

Sea fennel leaves (*Crithmum martimum*) have been used as a medicinal herb, a spice and a salad ingredient in Europe and Asia where species of sea fennel grow wild in the dunes. Traditionally sea fennel was gathered and eaten by sailors as a protection against scurvy. Local peoples gathered sea fennel for its fresh sharp taste. It contains high concentrations of Vitamin C and has been compared in taste to kale and arugala.

Local Importance: Wild crop, medicinal.

Potential Importance: Vitamin C source for poorest and driest areas.

Sea Purslane: A Tasty Mat of Vegetation

Sea purslane *(Sesuvium portulacastrum)* is cultivated in Indonesia, India and southern China. There it is consumed as a vegetable after being boiled in several changes of water until it is tender and most of the excess salt has leached away. This plant also grows wild on the Caribbean coasts and the coastal areas of southern United States. It is rich in calcium, iron and carotene. In India sea purslane is fed to animals.

Local Importance: Wild crop, cultivated vegetable, potherb, fodder

Potential Importance: Possible green crop for coasts.

Atriplex: Triangular Spinach

Atriplex (*Atriplex triangularis*) is an annual spinach like plant of the salt marshes of North America. Some work has been done in cultivating and selecting strains that yield over two tons per hectare with salt water irrigation. A related species is cultivated as a vegetable India. The leaves are eaten fresh or boiled.

Local Importance: Minor vegetable crop. Wild crop, source of leaf protein.

Potential Importance: New vegetable crops for the poorest coastland saline areas.

Common Purslane: Vitamin C in the Garden

Purslane (*Portulacae vulgaris*) is another salad plant with a surprising amount of tolerance for salt. A spreading tender mat of stems and bright small leafed foliage it can be consumed as a potherb in salads and soups or fed to animals. It is especially palatable to birds and poultry. Common purslane has long been known as a plant which can cure scurvy. Since purslane contains 29 mg of Vitamin C per 100 grams and a Vitamin A potency of 7,500 IU per 100 grams, this tradition has been confirmed by scientific investigation.

Local Importance: Potherb, salad plant, animal food.

Potential Importance: Reclamative plant, new minor vegetable and fodder plant.

Alkalai Sacaton: Food Resource of the Southwest US

The Hopi and Piute Indians gathered the grain from this perennial grass (*Sporobolus airoides)* which grows wild in the western US and Mexico. The approximately one mm grain is easily gathered by the use of harvesting baskets. It is abundant and separates easily from the seed heads. Other sacaton varieties also grow on saline and alkali soils but their value as food plants have not even been investigated.

Local Importance: Lost crop, reclamative, anti-wind erosion.

Potential Importance: New family of cereal crop candidates for arid lands.

The Bounty from Bitter Waters

Each of these plants can provide excellent nutrition. Many of them are tastier than food bought in the supermarket. Most importantly their cultivation would not involve scarce fresh water and could be done on marginal lands. Certainly these palatable halophytes are deserving of serious research and consideration as we strive to achieve abundance.

IX. Strange Food from Stranger Places

Food to sustain human life can be found in many of the extreme environments of the world. In Laurens van der Posts's book *The Lost World Of the Kalahari* the author describes the bounty these hunter-gatherers harvested from the sandy, arid apparently foodless wastes, finding enough nutritious food in a few hours to nourish them until the next day. (van der Post, 1967, pg. 31)

The depths of Sahara, such a fearsome place that few dare to cross it even with the aid of four wheeled vehicles, was once an area from which grain was *exported.*

Other extreme environments also harbor botanical treasures. Many of these plants are still harvested and eaten by the rugged inhabitants on the fringes of the inhabitable world who find in them the concentrated nutrients and energy needed to survive.

Are there plants in these places that might "come out" into the world and take their places as tomorrow's crops? Could identifying these foods and learning about these foods help us toward true abundance?

The Mustard Caper from the Sinai Peninsula: Bounty of the Wadis.

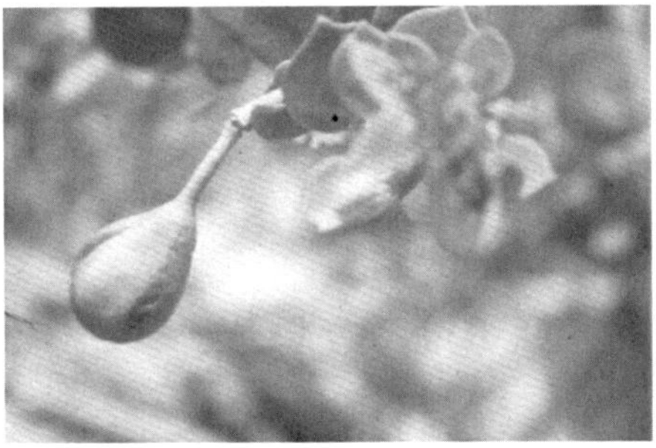

Photo by Elaine Solowey

Figure 7: Mustard Caper

In the wadis between the bare reddish peaks of the Sinai the tourist or hiker sometimes comes upon a startling sight. A colony of plants with intensely green fleshy leaves, creamy orchid like flowers and lush purple fruits will have spread across a dry canyon or grown out of the solid rock.

Wild bees, birds and antelopes are making the most of this unexpected bounty and flee at the sight of an approaching human. If the wanderer picks one of the fruits, 50 to 100 grams in weight, looking like a smooth purple elongated fig, he will be tempted to eat it despite the sharp mustard like scent. Surely a fruit so moist and beautiful must be good to eat.

This sharp scent, which distinguishes it from its domesticated cousins, is actually a warning. The beautiful fruit should be cooked or fermented before ingestion by people.

This is the mustard caper (*Caperis sinaca*), a spiny deciduous plant with a spreading habit. It is capable of rooting in shallow stony soils and sometimes growing out of cliffs or over stone walls. It is a close relative of the Mediterranean caper bush (Caperis spinosa) which is esteemed for its medicinal properties and spicy edible buds.

The mustard caper is more drought tolerant and salt tolerant than the common caper and vigorous in arid and saline areas. Unlike the common caper its buds develop into a fruit large enough and sweet enough to be made into jelly or wine.

Capers need small amounts of water and well rotted manure to produce a crop. The mustard caper does not even need these small inputs growing out of the reddish rock walls of the wadis or covering areas where water has sunk into the earth sometimes months before the caper seeds sprout.

Currently the status of the mustard caper is that of a wild crop. Fruit is gathered by local Bedouin on the Egyptian side of the border and there is some jam and wine-making going on in Israel where the mustard caper is esteemed as a delicacy and a medicinal plant.

Certainly a plant which can survive and produce fruit on 50 mm of rainfall should not be ignored. The mustard caper's value to conventional agriculture may be in its ability to pass on its drought tolerance and salt tolerance to its domesticated relatives. But the hiker in the desert sees the beauty of the plant where among the saw-toothed mountains and dry black ravines of the Sinai the caper plant is like an unexpected gift.

Maca Roots from the Puna: Butterscotch at the Top of the World

Maca (*Lepidium meyenii*) is found at higher altitudes than any food crop in the world. It grows 4,500 meters above sea level in northern Peru's Puna region: a chilly, inhospitable treeless area where temperatures may fluctuate from 20°C to -10°C in a single day. The sunlight is intense at this altitude and the air is fiercely dry. Freezing winds sweep exposed rocky areas and the patches of arable soils are few and far between. Even the bitter potato cannot grow in the Puna.

That mat-like maca plant not only survives but produces tasty roots rich in sugars, starches, proteins, iron and iodine.

The maca plant is one which could easily go unnoticed, lying flat on the cold, stony ground. No part of the plant is wasted. The leaves are used to fatten guinea pigs for the table. The roots can be boiled or baked and contain between 13 and 17 percent protein.

Maca can be used fresh or dried and kept for years in storage with little nutrient loss. Maca can be boiled in water to make a butterscotch like drink or cooked into a spicy pudding. The roots have good effects on the fertility of animals and humans. After the Spanish conquest ailing colonial breeding stock, unable to adjust to the altitude, were fed on maca roots to improve their viability.

Cultivation of the maca is complicated involving the selection of roots, transplanting the roots into a manured paddock, allowing them to set seed, turning sheep or llamas into the paddock to tread the seeds in, then mulching and lining the maca field with stones. The results are extraordinary. From the desolate frozen earth comes large radish like roots in purple and yellow that contain all the good things that are needed to sustain life in the Puna.

As a poor people's crop maca was never investigated seriously by botanists or agronomists until the 1980s. Peasant farmers were encouraged to forget maca and grow corn at lower elevations. This lead to a steady decline in the number of acres planted. Maca was declared a crop plant in danger of extinction in 1982 and its future is by no means assured though scientists have become belatedly interested in its effects on fertility and its high nutritive value.

Still maca survives to this day in tiny scattered farming plots at the top of the world. Its wild ancestor survives at even higher altitudes. Both are waiting for recognition and development.

Surely a food plant that can grow despite low temperatures, high winds, frozen earth and intense sunlight may be useful in other chilly desolate areas.

One can only hope that maca will survive and be of use in the fight to feed the hungry world.

The Marula Tree from the Kalahari: The Bushmen's Supermarket

The marula (*Sclerocarya caffra/bierra*) is a large tree with light red grained wood and blue green leaves that grows in the Kalahari, South Africa and as far north as Zimbabwe (colorplate 20 on page 267 and colorplate 21 on page 267). A striking feature of the dry bush and sandy wastes, the marula towers over the Kalahari's sparse vegetation, making islands of deep shade.

Once a year the marula drops copious amounts of small mango-like fruit. Hard and green when these fruits abcisse they ripen on the ground over the period of a week to a carpet of reddish gold fragrant morsels that can be squeezed into the mouth like little sacks of juice.

If not gathered by the indigenous people and made into wine, brandy and beer, the fruit ferments on the ground, giving off an overpoweringly sweet tangy scent that attracts animals from miles around. Elephants, gnus, rhinos and antelopes have gorged themselves on the fallen fruit and become intoxicated. One such occurrence was filmed for a popular movie and clearly shows the beasts under the influence of alcohol.

The marula fruit is something of a nutritional powerhouse containing large amounts of Vitamin C. The large seed of the marula can be pressed for an excellent non-drying oil which is both edible and valuable as a cosmetic or roasted and eaten as a nut.

Many tons of marula fruit are gathered in the wild and made into a liquor in South Africa and Botswana each year. Test orchards in Israel have proved the marula both salt tolerant and drought tolerant to a high degree. As a close relative of the mango and the cashew it is possible that the marula's hardiness may be useful in extending the range of these valuable cultivars.

Or the marula itself may soon may be planted as an orchard tree. The variability of fruit type and size, amount of sugars present in the fruit, the high yield, immunity to pests and low water requirements all make the marula an attractive crop candidate.

It is entirely possible that within the next decade marula products will be sold in the supermarket, called perhaps "Health Food from the Kalahari."

Table 4: Food Value of the Marula

Marula Sclerocarya caffra/bierra Food value per 100 gram of edible pulp

Calories	50-55
Moisture	85-86
Protein	0.5-0.6
Fat	0.4
Carbohydrates	12-14
Acids (citric)	8-12 g
Fiber	1.2-2 g
Ash	0.9-1.3
Magnesium	30-70 mg
Calcium	20-120 mg.
Phosphorus	8-12 mg.
Iron	0.3-0.5
Sodium	2.0-2.5
Potassium	163-490 mg
Nicotinic acid	0.27-0.48 mg.
Vitamin C	130-270 mg.

Analysis by Ben-Gurion University

Drinn: The Black Gift of the Sahara

In previous centuries grains of the deserts and savannas were harvested in enormous quantities. In the Sahara a household might harvest a thousand kilos of seeds from a wild grass stand.

The seeds were piled up at caravansary and shipped out of the Sahara and the Sahel by the camel-load. This was a major source of income for the tribes of these dry searingly hot areas and has no modern equivalent.

The mixture of wild cereals, called **kreb** was not "famine food" but a most esteemed treat as far away as Cairo where it was purchased by the wealthier citizens of the city.

The wild cereal harvests of these areas have been in decline for a century, reflecting the decline of the nomadic people who tended and harvested the immense wild grass stands. Sedentary life means over grazing and many wild grasses simply do not survive long enough to set seeds.

The sustainable practices behind this ancient harvesting system have also been forgotten. At one time the tribal chiefs and elders were responsible for allotting grazing rights and could slaughter animals found in wild grass stands reserved for human use.

Just because the wild cereals of the waste-lands are not important food crops now does not means that they will not play an important role in the future. Evolving in the worst and harshest of conditions these hardy plants could be exactly what are needed to preserve the fragile desert ecosystems and still provide the remaining indigenous people with food and income.

Many wild cereals are perennials; they do not have to be replanted year after year and the seed that would have to be saved for replanting can be eaten. Perennial grasses also provide continuous protection against wind erosion and water loss.

The hardiest and most abundant of the these grains is drinn, (*Aristada pungens*) "nomad rice," "ebon corn," the black gift of the Sahara.

Drinn was by far the most important grain of the northern Sahara growing on sand dunes and in low areas watered by run off. It is a tall tufted perennial with deep roots, long leaves and black grains.

Hillocks of sand on the Saharan plains are still covered with enormous tufts of drinn the edible grain found at the tips of long stems that sweep the sand.

Travelers crossing the Sahara wrote about the value of drinn for food and forage. Duveyrier, an explorer, commented in 1864 that drinn was the only food available for people and animals (BOSTID, 1996, pg. 260).

As recently as 1969 drinn was still an important part of the diet of nomad tribes in the area.

Can drinn be useful in the modern world? A nutritious grain which grows in the blazing sand and thrives on less than 30 mm of rainfall is a promising crop candidate for arid and saline areas.

Drinn seed is already as large and palatable as some conventional grains. With minimum investment in selection, collection and agronomy, drinn could take its place as a major cereal crop.

Nuñas: Popping Beans from the Uplands of Ecuador and Peru

Beans are an important protein food in Peru and Ecuador (colorplate 15 on page 62). But beans are sometimes difficult to cook and even common varieties take hours to cook properly. In the treeless high plains, in areas too desolate and sparse to do more than graze sheep and llamas, water boils before it is hot enough to make tea and fuel is scarce. So how is it possible to cook a pot of beans?

The answer to this problem is the nuña, a type of common bean that is the bean counterpart of popcorn. When heated the beans burst out of the seed coat with a soft hiss opening up into a butterfly shape, becoming soft and tasty with a flavor like roasted peanuts.

This little known variation of *Phaeseolus vulgaris* is grown above 2,500 meters. They look like common beans and come in striking shades of white, red and black but they are exceptionally hard shelled.

When cooked the heat and moisture build up inside the shell, then cause the bean to burst. This usually happens in less than five minutes. Popping nuñas requires much less fuel than boiling ordinary beans even if it were possible at such heights. The nuñas are simply heated in a little oil like popcorn and then begin to open into a delicious palatable source of high quality protein.

Nuñas are usually grown for home consumption. They are commonly toasted in a hot frying pan coated with oil or animal fat. The toasted nuñas are then served as a side dish. In southern Peru they are sometimes sold to tourists as an authentic local food. Very little of the crop ever comes to market.

There is some evidence that nuñas have day length sensitivities and can currently be grown only around the equator though their other cultivation needs are minimal. Like most beans in small farming operations the nuñas are usually grown beside upland corn and allowed to use the corn stalks as a trellis.

Many other day length sensitive plants, including the common potato have successfully adapted to other latitudes. Nuñas could be a nutritious snack for the developed world, replacing snack foods high in sodium and fat and low in food value. For the developing world nuñas could be an excellent source of high quality protein that takes very little energy to prepare.

The Morama Bean: Two-Tiered Treat from Namibia

One of the most amazing of the moisture-storing plants from the deserts of southern Africa is the morama bean (*Tylosema esculentum*). Native to Namibia, the sandy areas of South Africa and the Kalahari, the morama bean produces two edible components: seeds and tubers. It is a large trailing shrub with bilobed leaves and yellow flowers which grows out of a very large underground tuber. The yellow flowers quickly turn into pods. The tuber supports the plant during the dry season which can last up to eight months.

Mature morama plants can survive for up to five years without water, using only what is stored in the plant's tuber. The morama tuber can weigh up to 30 kilos. The largest specimen ever found was said to have weighed 250 kilos but such specimens are usually too old and tough to eat. Young tubers are much more palatable, growing to the size of a large potato in dry years. In wet years the tubers swell quickly and may obtain a weight of three kilograms within a few months.

When cooked the tubers have a slightly sweetish taste like artichoke hearts and contain 9 percent protein.

The pods of the plant contain between two and six large dark brown slightly slimy seeds. They are tasteless raw but when cooked become as tasty as the cashew nut.

The seeds have a protein content of 30 to 39 percent which is the equal of the soybean. The oil content of the seed is about 40 percent. Seeds store well and are remain edible for months.

The seeds germinate after two to six days in encouraging conditions and apparently do not have a dormancy period or contain growth inhibitors. This is rare for a true desert plant and could be important if the morama becomes a cultivated crop.

Experiments in Texas at the Chihuahuan Desert Research Institute in the 1980s proved that the morama bean develops rapidly into an edible crop if planted in sandy soil and irrigated lightly with a drip system. Vegetative growth continued through the hottest months with no dormancy. Tubers from the irrigated test plots gained a weight of two kilograms.

Tubers or nuts? It is obvious that the morama could become a tasty and nutritious cultivar when we finally get around to deciding what we want from the plant.

Either component contains the nutritive value of some of our most esteemed cultivated crops and can be cultivated with minimal inputs on the least promising of marginal lands.

Quandong: A Surprise from the Australian Bush

A quandong tree, growing out of the dry soil and glowing with ruby-red fruits is an amazing sight (colorplate 22 on page 268).

This arid zone tree, which has long leathery leaves like an acacia and a brightly colored peach like fruit that resembles an edible Christmas tree ornament, is considered one of the few Australian native plants with potential horticultural and economic importance. (Another is the macadamia nut.)

The quandong (*Santalum acuminatum*) has long been recognized as nutritious food. The Australian Aborigines dried and stored the flesh for future use and also ate the kernel.

The flesh is high in Vitamin C, containing roughly twice the amount found in oranges. The kernel yields approximately 70 percent oil and 20 percent protein.

Quandongs produce fruit under arid conditions with little rainfall. There is some evidence that quandongs regularly parasitize other plants, so planting the quandong with a nurse tree in close proximity usually guarantees the success of the quandong sapling.

The flesh of the quandong is thin and tastes much like an acerola but there is great variability in the wild stocks and the quandong could improve rapidly with minimal attention.

The Commonwealth Scientific Industrial Research Organization (CSIRO) Division of Horticultural Research in Australia has been investigating the quandong's horticultural potential since 1973. Researchers have discovered dozens of fruit types and colors, from the perfect round red varieties to pear shaped yellow types with thicker flesh and small kernels. Techniques of seed germinating, grafting, root-stock and scion selection have been developed for the quandong. Well established trees are copious bearers. Young trees irrigated with saline water produced 1,600 fruits. Older trees have produced 4,000 well formed fruits with saline irrigation.

The fruit sells well in the cities of Australia despite the high prices of fruit grown in small experimental orchards or gathered in the wild. The quandong fruit is often made into jam or pies. It is served as stewed fruit or minced into kamaradin (fruit leather). Even fresh it is quite acidic with an odd aftereffect: it is very unpleasant to smoke a cigarette for several hours after eating a quandong because of a powerful and unusual aftertaste.

Brian Powell of Quorn was one of the pioneers who set up the first quandong orchards under the auspices of the CSIRO in 1975. He sold 180,000 fruits in 1986. He commented in the same year *There's nothing to stop anyone from going ahead with it now on a commercial basis.* (Powell, 1986, pg 7)

Since then more orchards have been established and it seems that the quandong will become the second indigenous Australian plant to join the ranks of cultivated crops.

Monkey Orange: Lost World Fruit

Photo by Fritz Geral Schroder (Reprinted from The Growing Edge)

Figure 8: Monkey Orange

The monkey orange has absolutely nothing to do with monkeys or oranges. It is one of the strangest fruits in the world and is found in truly forbidding wastes that are mostly uninhabited to this day.

Encompassing three types of strychnos fruits (*Strychnos pungens*, *Strychnos spinosa* and *Strychnos cocoloides*) the monkey orange is said to have been placed by God in the wastes of southern Africa to provide food and drink for those who are lost. The charming legend was related to missionaries by the !Kuh bushmen.

Growing on sandy granite slopes in Natal, Namibia, Swaziland and Mozambique the strychnos trees are small deciduous and spiny with sparse branches. Trees are able to survive and bear fruit with less than 100 mm yearly rainfall. The bark is rough, pale gray or white. The leaves are thin and glossy on the upper surface. The flowers are borne in clusters at the end of the spiny twigs. That is where the fruit develops, large round balls as hard as wood that have to be broken open like coconuts.

When green the monkey orange does not look edible. Ripe, the fruit gives off the odor of cloves, the hard rind becomes slightly yellow and the pulp inside has a pleasant tangy flavor resembling a mixture of pineapple and orange. The pulp is found in flaps around large smooth seeds. The seeds are said to be inedible and are discarded by the bushmen when the fruit is eaten. Interestingly enough baboons also discard the seeds of the monkey orange though they crack the fruit with their powerful jaws and avidly eat the pulp.

The "shelf life" of these fruits is just short of fantastic. They keep without refrigeration for several months and if buried in the sand are still good after several *years.*

It is not known if the strychnos fruits have any economic potential. Some plants have made their appearance in botanical collections as the "wood-orange."

Saplings brought to Israel and planted in the Arava desert bore fruit after three years so the trees can adapt to other arid and saline areas.

Very little research has been done on the monkey orange's food value or its horticultural requirements.

Because of the extreme variability of the types, tastes and sizes of the wild fruits a rudimentary selection program may discover varieties which can be easily domesticated. or collected in the wild.

Until then the monkey orange will continue to be food to the bushmen and animals and solace to the lost, planted in the middle of nowhere by the grace of God.

Bourgou: Food from the Lakes of Grass

The inner delta of Mali is underwater several months of the year until the floodplains dry out. In areas where vegetation does not grow the earth dries into a cement like surface marred by deep cracks. But in areas where the bourgou (*Echinochloa stagnina*) grows a dense mat of grass a meter thick preserves the soil and holds in the moisture.

The bourgou grass has a strange life cycle. Trodden in by livestock as seeds and runners when the area is used as temporary pasture, the grass sets down deep roots which survive when the area is grazed and later dries out.

When the wet season comes the deep roots put out shoots that grow quickly as the water rises. In the second month of the floods the grass flowers. After three months the bourgou is three meters tall and the flood plain is approximately two and a half meters under water. The bourgou then sets seed. As the water level falls the area is used as a temporary pasture again and the mat of grass will sustain the livestock over the dry season.

Bourgou grass can be gathered and sold as fodder. But some people who live in the vicinity of the bourgou swamps gather the seeds for food and cut and press the young grass for sugar.

Bourgou is related to two other wild grains of potential importance; antelope grass and shama millet. These also are grown on swampy, clay soils with fluctuating water supplies. Antelope grass is used primarily as fodder but shama millet was cultivated by the ancient Egyptians as a cereal crop.

An intense period of drought from 1968 to 1985 destroyed many bourgou swamps and some areas formerly covered with bourgou have been converted to rice cultivation.

The government of Mali and the United Nations Sudano-Sahelian Office (UNSO) embarked on an ambitious project to restore the bourgou swamps and encourage the unusual and sustainable land usages of the area. By 1990 4,000 hectares had been replanted by hand with bourgou grass and farmers were able to sell fodder, extract sugar and sustain their livestock with the restored bounty of the "lakes of grass."

The Domim: Fruit Tree of the Arava

The Arava is a dry, hot, narrow barren valley that runs from the southern tip of the Dead Sea to the Red Sea. Temperatures vary from minus three C in the winter to a sizzling 50 degrees C in the long rainless summer. Winter rainfall is rare but still causes flooding as there is little vegetation or soil to absorb the water.

The border between Jordan and Israel follows the Arava for over two hundred kilometers and the colorful desert landscape of red rocks, white sands and blackish purple mountains is often scarred by barbed wire fences and old mine fields.

This does not seem to be a likely place to find fruit trees loaded with sweet red fruits but the domim tree (*Zisiphus spinichristi*) also called Christ's thorn graces the forbidding landscape with green leaves, white flowers and candy-like fruits.

Related to the jujube (*Zisiphus jujube*) and the Indian date (*Zisiphus mauritania*) the domim tree is a short, spreading thorny tree with multiple trunks. Domim grow in wadis and other dry watercourses, using the moisture that sinks into the sand during the infrequent floods. In upland areas the domim trees sprout in ditches and creeks. The fruits are generally small, date like when dried but more like an apple when eaten ripe.

Domim are a favorite wildcrop in Israel eaten with relish by hikers, soldiers, tourists and often gathered to be dried or made into jam.

The fruit is eaten by the Beduin in Israel, Jordan and the Sinai peninsula. The dried fallen fruits are a favorite browse of deer and antelopes. Some southern domim types have flowers that stink, giving off an odor that is very much like carrion, probably to attract the needed pollinating insects. Pink and white fragrant blossoms are more likely to be found on trees that grow in less arid areas.

The close relationship between the domim tree and two minor domesticated crops hints at potential horticultural value.

For the sojourner in the forbidding Middle Eastern wastes the domim tree is a pleasant and welcome surprise.

Uvilla: Tonic Grapes of the Amazon

The uvilla (*Pourouma cecropiaefolia*) is found in the rainforest of the western Amazon.

People who live in the native range of the uvilla have difficulty finding or growing enough food. The rainforest is lush in this area but is not made up of many edible species. Areas cleared for raising conventional crops are often swallowed up by jungle regrowth in less than a season.

Few fruits grow in the wet equatorial forests but the uvilla is prolific and fruits heavily during the entire three months of the wettest season.

Uvillas grow on small broad leafed trees that mature quickly and begin to fruit in their third year. Male and female flowers are borne on separate trees. The trees sprout easily from seeds and have no known natural enemies.

The large purple grape like fruits contain a single large white pit and can be eaten raw or pressed for wine.

Indigenous people collect the fruit from wild trees and cultivate uvillas as dooryard trees. Wines and preserves made from uvillas are consumed in the home and sometimes sold in local marketplaces.

Uvillas have value for home planting or as components of sustainable land use systems in the humid tropics. Little is known about the possible uses of the fruit as flavoring or its ultimate horticultural potential.

Two characteristics make the uvilla stand out from the many related species of Pourouma that grow in the rainforests and also bear reputedly edible fruit.

The uvilla is apparently impervious to pests.

Many local peoples swear that eating the uvilla is the only thing that keeps them healthy during he wet season.

It might be *very* interesting to find out why.

Yehib: A Unique Plant from the Horn of Africa.

The yehib (*Cordeuxia edulis*) has become very rare. A single grove exists in Voi, Kenya. Destroyed by years of war and famine in their native range a dwindling number of trees remain in the wild in the Horn of Africa.

The yehib produces a delicious edible nut which can be peeled with the fingers. The short round fleshy leaves of the yehib exude a brilliant red dye. The leaves and roots are purported to have medicinal properties.

The usefulness of the yehib may have proven to be its undoing. In times of famine every single nut was collected and eaten. The branches were lopped to feed livestock and sometimes the root was dug up to be pounded into medicines for a people cut off from medical care.

Obtaining yehib seeds is difficult and transplanting the trees to other countries in order to preserve them has not met with much success. The tiny yehib seedling quickly develops an immensely long tap root which apparently must grow straight down. This means that the yehibs must be planted out before the root outgrows the pot.

Such tiny seedlings do not always survive. Even larger trees are not cold hardy. In the southern Israel yehib trees four years old were killed by a few hours of low temperatures despite the fact that the trees had been wrapped in insulating material.

What makes this plant worthy of further cultivation and investigation? The yehib is a leguminous nut, a plant so odd that it has been classified as the only species of a new family of plants. It survives and produces high protein food in the forbidding areas of the Horn of Africa and the Ogden desert. Indeed in some areas it is a keystone species, underpinning all the plant and animal life of truly desolate plains.

Perhaps some absent soil symbiont or missing nutrient has prevented its successful transfer to other arid lands of the world.

Letting such a plant become extinct before we have any idea of its potential is foolish and irresponsible from both the ecological and agricultural viewpoint.

This is something we cannot afford to do if we are ever to achieve true abundance.

X. Food From Under the Earth

Many of the local staple crops of the world are tubers, thickened underground stems which store food energy for the plants. Tubers are generally less nutritious than seeds because they have higher water content and usually contain less protein.

However from a cultivator's point of view tubers can supply more food energy per hectare than fields of grain. Many tubers grow on soil unsuitable for grain crops and produce food under difficult conditions. Tubers are easy to cultivate. After being planted in a prepared bed little work is required except banking and harvesting.

Improving existing tuber crops or discovering new ones would greatly improve world food security and bring us one step closer to abundance.

The undisputed king of the tubers is the "common" potato, the most useful and popular food plant after corn, wheat and rice with a worldwide production of over 300 million tons per year.

The common potato (*Solanum tuberosum*) is a direct descendent of the potatoes brought from South America to Europe hundreds of years ago. These plants were of the species *S. andigena*, only one of seven distinct races of potatoes grown in the Andes.

Potatoes were not an immediate success in Europe. There was suspicion about a food not mentioned in the Bible. Many plants failed to adapt to different day lengths and climates. In the 19th century the potato was thought to be food fit only for peasants, lazy degraded ones at that. This notion is not readily understood by the potato chip loving modern world but Van Gogh's painting *The Potato Eaters* is a clear illustration. The work depicts the spud-eating farmers as dull, dirty and ugly serfs.

Potatoes became more popular and important as the efforts of amateur horticulturists produced plants more suitable to Europe. The potato also had champions among the royalty and intelligentsia of the day. It was found that the potato would grow in cool, wet areas unsuitable for wheat cultivation. The crop that had formerly sustained the population in these areas was rye. But rye was often infested with a fungus (ergot) that caused periodic outbreaks of insanity, convulsions and nausea among people who lived on rye bread.

In these areas the potato quickly became the staple crop despite the absurd prejudices with which it was regarded. Indeed, in some places potatoes were virtually the only food available, i.e. Western Scotland and Ireland in the 19th century.

Some of these prejudices lived on into the twentieth century and the potato was considered fattening stodgy food. Recently the true value of the potato has become known. Potatoes provide much of the Vitamin C of northern diets and are a true storehouse of many substances needed to sustain human life. But controversy still rages about these tubers with other food experts claiming that potatoes raise the blood sugar as much as candy or white bread.

Potatoes have become global favorites when sliced and fried as chips. They are recommended baked or boiled as diet food and appear in the cuisine of most countries in casseroles, potato pancakes, dumplings, soups, mashes and stews. Potatoes can be made into flour, processed into weaning food and turned into alcohol as well.

The one species of potato that entered the agricultural mainstream produces an amazing array of flavors, colors and textures. *Solanum tuberosum* can be waxy, floury, crisp, sweet, nutty, acidic and colored white, red, yellow and brown.

But what about the other potatoes? There are 154 known species of potato, seven **races** that are cultivated in the Andes. Within these species is a truly astonishing richness and diversity that could be harnessed to help feed the world.

The Potatoes that Stayed Home

The descendants of the andigena potato are cultivated in 150 countries. But the andigena potato is by no means the hardiest, the most nutritious or the best tasting potato.

The potatoes that never entered mainstream agriculture may be helpful in the coming struggle to feed the hungry world. Here is a brief description of some of the species found in the Andes and an assessment of their potential value.

Pitiquinia

This species (*Solanum stentomum*) produces long, cylindrical, knobby potatoes in red, black or white with deep eyes. Some are twisted or in spiral shapes. Pitiquinias are unusually high in Vitamin C, protein and solids. They are esteemed for their pleasant nutty flavor. This is a diploid species considered among the most primitive of the potato species. Pitiquinias are frost tolerant and produce fertile seeds but are considered rare and not grown outside of the Andes.

Local Importance: Delicacy and regional crop.

Potential Importance: Possible crop for slopes and uplands with poor soil, carries traits that may improve the common potato.

Limena

These potatoes (*Solanum goniocalyx*) are also known as "yellow potatoes." This species has round tubers, yellow flesh with deep eyes and exceptionally good flavor. It is especially tasty when fried or put in soups. Limenas are widely grown in the temperate parts of the Andes and bring good prices in the marketplace. It is not a particularly frost resistant species but it is resistant to disease. Being closely related to the pitiquinia potato the limena is high in protein and Vitamin C.

Local Importance: Local staple and cash crop.

Potential Importance: A species potentially equal to the "common" potato.

Phureja

These potatoes (*Solanum phureja*) are small, irregular in shape, purple and brown in color with a firmer texture and a stronger flavor than most potatoes. The deep-eyed gnarled phurejas do not have a dormancy period and are resistant to diseases and pests. These qualities make this species suitable for warmer climates. The extreme biodiversity within the species has already proved valuable. Genes from two phureja potato strains have been transferred to two heat tolerant varieties of the common potato. One type has become popular in the Netherlands because of its resistance to pests.

Local Importance: A potato for warmer areas.

Potential Importance: Already improving existing potato strains, a new crop for areas where the common potato is failing.

Chaucha

These potatoes (*Solanum chaucha*) are sterile triploid plants that produce no seed and have to be propagated by seed potatoes or eyes. The chauchas are thought to be the results of hybridization between the andigena potatoes and the pitiquinias. The chauchas combine the easy cultivation needs and the large firm tubers of the common potato with the nuttiness of the pitiquina. Chauchas tend to be crisp with flesh that can be white, yellow or brown. In Peru one stain called the huayro, has spread from the highlands and replaced commercial varieties that do not taste as good or yield as well.

Local Importance: Staple and cash crop.

Potential Importance: Frost tolerance, vigor and hardiness traits to improve existing potato lines.

Ajanhuirl

This high altitude race (*Solanum ajanhuiri*), most of them blue or bluish have a sweet floury taste (colorplate 24 on page 268). Some are palatable without cooking. Others are bitter and are freeze-dried to make the easily storable "chuno" staple. The tubers tend to be long and round with shallow eyes and contain generous amounts of Vitamin C and dry matter. Blue potatoes mature early and can withstand freezing temperatures, hail, drought and water logging. Blue potatoes are resistant to nematodes and most viruses.

Local Importance: Dried mountain staple.

Potential Importance: Crop for cold, stony areas with poor drainage. Source of cold and disease resistance for common potato lines.

Rucki

Rucki potatoes (*Solanum juzepczukii/curtilobum*) are the most frost resistant and even tolerate frost during the growing season. They can be grown above 4,200 meters where frost can occur 300 nights in a year. Rucki potatoes are thought to be the result of hybridization between *Solanum x juzepc-*

zukii and *Solanum x curtilobum* and a tiny wild species found at altitudes where the earth never thaws called *Solanum acule*. Rucki potatoes were cultivated in the mountains before the rise of the Inca empire and are considered a safety net crop that produces when all other crops fail. The tubers are usually bitter and small, high in solids and suitable only for chuno production as they are too bitter to eat unless frozen and dried. Yields are low but modern selection and agronomy could improve Rucki potatoes in a short time.

Local Importance: Safety crop, dried staple.

Potential Importance: A potato for extreme environments, source of genes for cold resistance, disease and pest resistance

---

Wild Potatoes

There are hundreds of wild species of potato in South America some of which are virtually immune to the most formidable of pests including the Colorado potato beetle, the aphid, the leafhopper and the termite. Potatoes immune to blight also exist in the wild. Most wild potatoes are not very palatable but the genes they carry could be used to improve all existing strains.

Local Importance: Only occasionally eaten, food for hard times. There is apparently some gene flow between the wild potatoes and the Rucki and Ajanhuiri species.

Potential Importance: Pest resistance for the fiercest of the potato pests. Resistance to the resurgent potato blight. Ecologically sound alternative to genetically engineered pest resistance.

These important plants have been neglected for too long. In a rapidly changing world any plant that can be used to improve an chief staple like the common potato is a treasure. The idea that there are other potatoes "out there" just as good or better than the common spud is very exciting.

In addition to the forgotten potatoes of the Andes, the area of the former Inca empire is teeming with other tuberous nutritional bonanzas that should be investigated as soon as possible.

Achira: Lily with a Future

The achira (*Canna edulis*) is related to the large leafed ornamental canna lilies that are sold as garden plants in much of the temperate world and may have been one of the first crops domesticated by the people of the Andes. The big plants with beautiful foliage are sometimes used as living fences by indigenous farmers but are prized mostly for their tasty, starchy corms. Some corms are simply cooked and eaten but a large percentage of the crop is turned into a large grained digestible starch that is valuable in the food and textile industry. The starch is clear, glossy and superior to both corn and potato starch.

Achira is extremely adaptable and has been cultivated successfully in areas as diverse as the Caribbean islands, the hills of Indonesia and the north coast of Australia. It is a suitable intercrop for paddy culture and a promising reclamative crop for swampy areas.

Local Importance: Important safety net crop for Peru and Ecuador.

Potential Importance: Rice intercrop, fodder crop, vegetable and invalid food for the developing world.

Ahipa: Underground Bean

A relative of the pea, bean, soybean and peanut, the ahipa (*Pachyrhizus ahipa*) is one of the few legumes cultivated for its tubers. It is a small quick-growing plant from the Andes much more adaptable than many Andean cultivars because it is unaffected by differences in day length. Its crunchy delicious root resembles that of jicama, another root vegetable that has become a gourmet item as an addition to salads. The root may be steamed or stir-fried yet retains its crisp texture. When cooked its flavor can be compared to that of water chestnuts.

The ahipa tuber is a good source of potassium and Vitamin C. Its starch is easily digestible and may be suitable as diet food as it is low in sodium and contains just 50 calories per cup.

The ahipa vine does not climb but spirals around the tuber which is the only part of the plant suitable for human consumption. The plant is easily propagated by seeds or small tubers and requires little attention except for seasonal applications of manure. Tubers vary greatly in size and shape. Some tubers have brown skins and white flesh, others have flesh and skins of a magenta color.

Local Importance: Typical "lost crop" losing ground to conventional vegetables.

Potential Importance: Could be a new root crop for temperate areas. It is suitable for mechanical harvest.

Arracacha: The Carrot's Big Brother

The arracacha (*Arracacia xanthorrihiza*) was not given a botanical name until 300 years after the Spanish conquest. It was overlooked in the colonial period and never taken to Europe with other botanical riches such as maize or potatoes. Arracacha survived in small plots and upland gardens as a subsistence crop. Now it is gaining popularity in several South American countries including Brazil.

The two edible parts of the plant resemble celery and carrots. Underground the plant produces smooth carrot-like roots. Above ground the plant produces green celery like stems which can be boiled or eaten raw. The roots must be cooked. They can be added to soups or stews or boiled, baked and fried. The taste of the arrracacha root is something like a combination of cabbage, celery and roast chestnuts.

Local Importance: A comeback crop of good nutritive qualities and increasing commercial value.

Potential Importance: Arracacha should be tried in all tropical highlands in the poorer regions of the world. It may be a suitable cash crop for Southern Europe.

Cocoyams: Edible from Root to Crown

Cocoyams (*Xanthosoma sp.*) are native to South America but have already been introduced in other warm locales including West Africa. Humans or livestock can eat every part of the cocoyam. The large central corm can be chopped into fodder, the cormlets are suitable for human consumption, the tender shoots and leaves can be eaten like spinach and the woodier stems are a valuable green forage plant.

The cormlets have the most commercial potential being rich in vitamins, minerals and protein. Cocoyams are as palatable as sweet potatoes when baked. Corms can also be fried like chips or ground into flour. Edible corms are produced three months after planting and can be stored in cool, dry places for up to two years without spoiling or sprouting. The plant is easily propagated and tolerates waterlogging, swampiness, heat and humidity.

Local Importance: Important food crop for small farmers in some parts of Africa and tropical America.

Potential Importance: Possible protein/carbohydrate food for tropical, humid and wet areas around the world. May be important in animal husbandry.

Mashua: A Tough and Tasty Nastursium

Mashua (*Tropaeolum tuberosum*) is high yielding, delicious and resistant to cold and pests. Yet despite these qualities mashua is almost unknown outside of its native range. The tubers are about the size of small potatoes growing under a vine with beautiful flowers and lacelike leaves. Some mashua types have round roots, others are conical in shape. Eaten raw mashua roots are as spicy as radishes. When cooked the roots lose their sharpness and become sweet. High in protein and carbohydrates mashua roots are a local "famine breaker" that can be used to enrich soups or stews or can be fried and baked. In Bolivia and Peru mashua is even made into a popular dessert. Less labor intensive than potatoes, it is grown in the cool upland valleys of Argentina, Columbia, Bolivia and Peru.

Mashua is a classic "poor people's crop" fulfilling the role in the uplands that cassava does in the lowlands: a food which is unaffected by poor management and weather conditions that can be stored until needed. It is usually cultivated by women and children and reported to have antibiotic properties especially when eaten raw.

Local Importance: Mashua is often abandoned for more profitable crops but is still important to the upland poor.

Potential Importance: Mashua's antibiotic and pest resistant qualities warrant serious investigation. Mashua could be an important component in sustainable agricultural systems in the cool uplands of the world.

Makua: An Ancient Tuber Rediscovered by Modern Science

Makua (*Mirabilis expansa*) only became known to modern science as recently as the early 1960s when Bolivian scientist Julio Rea announced that he had found a new plant which was the staple food of a little known tribe of Indians. Following clues from folklore and legend he also found it growing in the higher uplands of Ecuador ten years later. In 1987 this mysterious culitivar was discovered in several locations in Peru. These locations are separated by hundreds of kilometers of rough terrain and any natural spread of the plant to these locales is unlikely.

A makua plant provides an abundance of juicy edible stems and a cluster of high protein tubers in chilly, dry, windy areas that place great strain on cultivated plants, including the hardier potatoes. The tubers are astringent when first harvested and must be exposed to the sun and wind until they sweeten or aged in a root cellar until the starch in the tubers hydrolyzes. In appearance the plant resembles a sweet potato though the gnarled makua tubers are reminiscent of the Jerusalem artichoke.

The stems and leaves can be eaten fresh or used as food for chickens and guinea pigs. Makua is richer than most root crops in calcium, phosphorus and potassium.

How did the makua come to grow in areas so similar in altitude and temperature but so widely separated by distance?

The answer could be in the Inca system of forced migration. Restless, newly conquered people were often moved to similar distant locales to keep them from rebelling against the empire. The Incan authorities were careful to choose locations in which captive groups would thrive and transplanted their useful plants and animals along with them.

According to legend people from the Cajamarca region were moved in this manner into the highlands of Ecuador and Peru many centuries before Columbus.

Local Importance: In the far-flung isolated areas where makua is grown it is very useful to people and animals as vegetable, fodder and root crop.

Potential Importance: Possible high yielding tuber for cold areas. Intercrop plant for upland corn, anti-erosive in cold windy areas.

Oca: Sweet Enough to be Fruits

Oca (*Oxalis tuberosa*) is appealing with stubby carrot like roots and firm shiny skins in warm colors ranging from hot pink to deep red. Oca tubers have been nicknamed the "rubies of the botanical world" by plant scientists who became interested in the oca in the 1960s. Most oca roots have a slightly acid taste, compared to "potatoes with sour cream" but some are sweet enough to eat like apples.

Bitter types, rich in oxalates are cultivated to be freeze-dried into a storable staple called caya or cavi. In the freeze-drying process the bitterness disappears.

Oca tubers are an excellent source of carbohydrates, calcium and iron and can be prepared in as many ways as the potato.

This compact hardy perennial plant is unfortunately photoperiod sensitive. It has been introduced to New Zealand in areas with similar day lengths. There it is cultivated commercially, harvested mechanically and called "the New Zealand yam." It has even become part of the national cuisine and is served with national dish of roast lamb.

Local Importance: In some areas oca is more important to the local economy than the potato.

Potential Importance: Oca could be an important crop for temperate and cold zones if day neutral varieties can be bred or discovered.

Taro: An Island Delicacy Comes into its Own

Taro (*Colocasia esculenta*) was little known in the United States and Europe until the late 1980s. It was grown widely in many locales and has become an important ingredient in the local cuisines of Hawaii, China and the Caribbean islands. Taro can be boiled, baked or fried in oil. It is primarily a carbohydrate food which has a mealy, pleasantly sweet flavor. Its easily digestible starch has been turned into baby food, and food for invalids and people with food allergies. Taro is an excellent source of Vitamins A and B, calcium, phosphorous and iron. In the developed world taro products made their first appearance in health-food stores.

Taro grows as a large corm with a particular affinity for marshy and waterlogged land. It can be integrated into paddy culture or grown on the edges of wetlands. If water is supplied at appropriate intervals, taro can be grown in drier areas. The large central corm is surrounded by cormlets, which are the most suitable for human food. Six to twelve months after the planting of a large corm the cormlets are mature enough to harvest. Taro is resistant to heat and humidity but can be vulnerable to pests. A search for pest resistant varieties, if successful, could greatly expand taro's usefulness and range.

Local Importance: Widely grown in the warm areas of the world as an fundamental part of the tropical diet.

Potential Importance: Alternative starch crop which could become a raw material for processed food.

Ulloco: Jewels from Poor Soils

The ullocos (*Ullucus tuberosus*) are even more brightly colored than the oca roots. The tubers are pink, red, yellow or even striped like candy. Their skins are waxy and shiny and they range in shape from small knobs to large crooked sausages. The texture of the cooked ulloco is crisp, silky and nutty. The roots do not have to be peeled. They are especially good in soups and stews because they absorb the flavor of spices and meat. The tubers have a good shelf life and the colors catch the shopper's eye in the urban or rural marketplace. Nutritionally the ulloco supplies Vitamin C, starches, protein and sugars. The tubers contain no fiber and no fat and are easily digested by most people.

Ullocos are propagated easily by pieces of stem or tuber. They are suitable for mechanical harvesting if the same care is afforded them as is taken with new potatoes. Tubers can be harvested after five months. While the ulloco does not have as high a yield as the common potato it is suitable for most temperate areas and could be introduced easily to zones with similar photoperiods.

Day neutral varieties of ulloco exist and some have been identified.

Local Importance: A much esteemed crop in its native range.

Potential Importance: A new root crop candidate for temperate zones.

Yacon: The Underground Sunflower

Yacon (*Polymnia sonchifolia*) is a sunflower relative grown for its edible tubers. The tubers are crunchy and full of moisture with a mild sweet flavor like watermelon. The tubers are usually eaten raw or cut up in salads. A sweet brown sugar called chancaca is made from the ripe tubers as well as a refreshing drink from the pressed, grated root. Unlike most other tubers the yacon stores carbohydrates in the form of inulin instead of starch making yacon a potential diet food as the human body does not metabolize inulin. The young stem of the plant is used as a cooked vegetable and the large veined leaves have promise as fodder. The plant grows fast and thrives in poor soils. It has already gained some popularity in the US as a garden plant and is grown in New Zealand as a specialty crop, packaged for sale like carrots.

Local Importance: A well liked minor crop in its native range, especially esteemed as holiday food.

Potential Importance: Sugar substitute, health and diet food. A new root crop for temperate and Mediterranean areas. A possible fodder crop which improves poor soils.

All of these splendid food plants have been ignored

by mainstream agriculture and are consequently underutilized. Learning to cultivate these new/old foods from under the earth is a wise step on the path to abundance.

XI. Edible Landscapes

The first farmers were probably nomads who harvested a crop, sowed some of the seed and then moved on, returning to the planted area after a year's wanderings. They lived in an edible landscape that supplied them with all their needs, so had no fixed abode. They simply had to gather, catch and prepare their foods from what grew around them.

Large groups of Native Americans continued to live this way until the middle of the 19th century, hunting, gathering and farming in a specific territory. It was an excellent way to support a small mobile population.

This lifestyle came to an end in North America as the land was settled by colonists from the overcrowded nations of Europe. Now there are very few people in the world who provide for themselves in this way. The wilderness is vanishing. Hunting and gathering cultures cannot survive in lands disturbed and degraded by human activity.

Fortunately, there are farmers today who amplify a natural system, then reap its bounty. These farmers still thrive in many parts of the world. Their systems are among the most productive and the most sustainable because they are managed in a way that uses the unique features of local ecosystems to best advantage.

The systems also provide employment for the local people and use their knowledge and familiarity with their surroundings as well as their desire to preserve their homes.

These systems fall roughly into three categories: traditional riparian systems, traditional agro forestry and modern integrated projects.

Dike/Pond System of Southeast China

An example of the first category would be the dike/pond systems of the Zhujiang Delta in Southern China. Seen from the air the delta is a lovely mosaic in shades of silver, blue and green, bisected by canals and a few roads. It is hard to believe this is one of the most densely populated areas of China, supporting the large economic center of Southeast China, the city of Guangzhou.

It is also one of the wettest areas in China. Several large rivers converge here including the Xijiang, Beinjiang and Dongjiang. The climate is humid with hot rainy summers, seasonal flooding from the monsoon rains and the influence of typhoons during the autumn.

Centuries of hard work, sound ecological principles and the need to support a large population have turned this steamy swamp into one of the most productive and diverse integrated systems in the world.

The core of the system is composed of three essential components: fish ponds, mulberry dikes and sugar cane dikes. These elements unite the other aspects of land use.

Fishponds are maintained by dredging and the soil and slime is used to repair or create dikes. The ponds are sanitized and fertilized by the use of quicklime, tea-seed cake and animal manure, all local inputs. River water is introduced to the ponds and the ponds are stocked with young fish. Meanwhile the pond mud has been used to fertilize and build up the upper surface of the dikes.

Mulberry trees and sugar cane are cultivated on the dikes but the dikes also support mushroom beds, gourds and melons grown on trellises to shade the ponds, groves of bamboo, fruit trees, stands of grass, vegetable gardens and rice paddies.

The mulberry trees support silkworm cultivation and paper making activities. Silkworm excrement, mulberry leaves, cocoon waste and dead larva are fed to the fish in the ponds.

Water birds and eels are raised in the pond water, as are aquatic plants of various sorts. Pigs are raised to provide manure and meat and fed with fallen fruit, cane waste, cut grass and food waste. Chickens are herded from dike to dike to eat insects. The bamboo is harvested to make roofs, baskets, fish traps, screen, trellises and frames, which are integral parts of the system.

The principle products of the dike/pond systems are: fish, duck and duck eggs, water hyacinths, water chestnuts, lotus, melons, gourds, eels, vegetables, rice, silk, mulberry fruits, other tree fruits, pork, piglets, sugar, chickens, eggs, bamboo, green fodder, hay, fuel and paper.

This is a truly amazing amount of useful and edible material to take out of what is basically a well-managed morass.

There is food everywhere, underfoot, in the water, providing shade overhead. The profitability of such a system is many times that of a conventional agriculture/aquaculture project and has the advantage of being sustainable. Many dike pond systems are hundreds of years old and show no decline in fertility.

A greater contrast between the teeming bounty of the dike/pond integration and the sterile monocropping of modern agriculture cannot be imagined.

Applicability: Other wet, monsoon soaked, typhoon lashed deltas exist all over Asia. Though agriculture and aquaculture remain the primary activity of the inhabitants no counterpart to the dike/pond system has developed. Instead many of these areas are being used for commercial aquaculture of the most damaging sort, notably shrimp farming which contaminates an area quickly and leaves it more impoverished and depleted than before.

Could the pond/dike system be transferred with some modifications to India, Indonesia, and Bangladesh?

The inputs needed to set up such a system are not expensive in material terms consisting mostly of mud, water, plants, animals and bamboo. Better flood control is one of the beneficial effects of the dike/pond integration. The sheer hard work needed to set up systems like these would be a blessing to the under employed laborers of those nations.

With skill, foresight, cooperation and a fraction of the funds wasted on commercial aquaculture dike/pond projects could turn intractable soggy regions into sustainable cornucopias of food and useful materials.

Floating Gardens: The Chinampa System

Chinampas provided much of the base for the might of the Aztec empire, which supported 25 million people on a diet of spirulina, fish, amaranth, maize, peas, beans, vegetables, turkeys and tropical fruits.

Chinampas do not actually float. They are platforms of mud piled up between post and vine walls. The mud is piled up and flattened out, then weeds and plant material are pressed into the upper layer. This is all covered with another layer of mud and fertilized with green algae and waterweeds skimmed off the canal.

The Aztec chinampas system started as a reclamation project to control the water around the ancient capital. It developed into an elaborate system of fertile gardens and shimmering canals around the empire's island seat of Tenochtitlan, which supported an unbelievably rich and diverse food-producing system.

Tenochtitlan was connected to firmer ground by a wide causeway. It was a city of approximately 300,000 just before the Spanish conquest with palaces, temples, pyramids, observatories and immense market squares where the agricultural wealth of the surrounding canals and chinampas was bought and sold.

Bernal Diaz who served with Cortez, commented on the quantities and wholesomeness of the food in his account of the Spanish conquest, *The Conquest of New Spain*. He mentions maize, small grains, domestic fowl, game animals, cocoa, tobacco, peas, beans, berries, tree fruits, prickly pears and herbs.

This is his impression of a market in Tlateclolo:

> *There were sellers of kidney beans and sage and other vegetables and herbs in another place and yet in another they were selling fowls and birds with great dewlaps (turkeys) also rabbits, hares, deer, young ducks, little dogs and other such creatures. Then there were the fruiterers and the women who sold cooked food, flour, honey cake and tripe.... I am forgetting the sellers of salt and the makers of flint knives and how they split them off from the stone itself and the fisherwoman and the men who sell small cakes made from a sort of weed which they get out on the great lake and forms a kind of bread which tastes rather like cheese. (Cohen, 1963, pg. 232)*

He says of a garden near Iztapala:

> *I was never tired of noticing the diversity of the trees and the various scents given off by each, and the paths choked with roses, other flowers and the many local fruits trees and the ponds of fresh water. Another remarkable thing was that large canoes could come into the garden from the lakes through the channel they had cut...everything was shining with lime and decorated with different kinds of stonework and painting which were a marvel to gaze on. There were many birds of many breeds and varieties. I say again that I stood looking at it and thought no land like it would ever be discovered in the whole world.... But today all I saw is overthrown and destroyed, nothing is left standing. (Cohen, 1963, pg. 215)*

The Aztec leaders were murdered, the waters around Tenochtitlan were drained away, the city was leveled, the gardens destroyed and the beautiful artwork buried in the mud. This was almost the end of the chinampas system. Isolated remnants remained in use mostly for growing vegetables and amaranth seedlings. Some chinampas are still in use today, having shown no decline in fertility in an estimated 2,000 years of use.

Applicability: An integrated system of great food producing potential, very well suited to intensive grain growing and aquaculture made basically from weeds, water, mud and the skill and dedication of the *chinampero.* This system could be set up in many wet areas of the world.

The chinampa system could also be an element for sustainable aquaculture systems.

Integrated Forest/Orchard Systems

Variations on this system are practiced in many parts of the world, usually involving the simultaneous planting of half a dozen crops in a cleared garden all that mature at different rates, preserve moisture, promote soil fertility and shade and support one another.

The garden is usually protected by a thick living windbreak, which also keeps out large animals and shaded by plantings of bushy or broad-leafed plants. Sometimes the garden is “roofed” by plants growing on trellis or shade screen woven from local plant materials.

The corn/beans/squash integrations typical of Mexico and South America are a good example. In a garden surrounded by useful trees and bushes a farmer will plant chickpeas, broad beans and peas. After a carefully timed interval corn, climbing beans and squash are planted. The chickpeas, broad beans and peas grow quickly and shade the small corn plants, squash and climbing beans.

The first group of plants set seed and are harvested and collected for fodder, revealing the second batch of seedlings. Those plants mature together, with the climbing beans using the corn as a trellis, the beans enhancing the soil fertility and the squash protecting the ground from the sun. Rocky areas and ends of rows are often planted with chilies, tomatoes or tomatillos.

The bushes and trees surrounding such a garden may be living fences like prickly pears, osage oranges, cactus apples and mesquite. Or they might be orchard plants such as white sapote, sapodillas and guavas.

The result is a seemingly untidy heap of vegetation that produces a dozen or more kinds of food and other useful products such as fuel or fiber from a small area with very little input.

Another variation is the tropical home garden or *shamba*. Home gardens in West Java may include over a hundred plant species which provide fruit, grain, vegetables, fiber, firewood, gum, spices, medicinal herbs, oil and tobacco.

Traditional gardens in Africa may include bananas, palms, tree fruits, yams, taro, cocoyams, pumpkins, squash, melons, millet, beans, and maize.

Traditional highland gardens in Sri Lanka may support over twenty species of trees as well as fifty kinds of annual plants that grow in their shelter.

A south American highland garden may be a paradisiacal collection of coffee trees, ice-cream beans, pitayas, lucumas, achiras and naranjillas.

These intriguing traditional methods are far more productive than modern monocropping because they produce such a wide range of products and they require so little external input. They are planted, managed ecosystems in miniature, fit carefully together by indigenous farmers not for monetary profit but to feed their families and provide for their needs.

Unfortunately these methods have all been lumped together under the heading of subsistence agriculture, which cannot compare in importance to "real" agriculture to today's agriculturalists.

What remains unsaid about subsistence agriculture is that it does little harm. There is little runoff of fertilizer or manure, no problem with herbicide or pesticide residues almost no erosion. The soil is enriched rather than compacted and most of the labor is done by hand or with simple tools and occasional use of an animal or a vehicle. Subsistence agricultural currently feeds more than one half of the world's population.

But even these time-tested and ecologically sound methods begin to break down when farmers are pushed into increasingly marginal lands by "real agriculture," which does not recognize the subsistence sector as anything but an archaic nuisance that must be swept away.

Applicability: Thankfully many of these systems remain in use and others can be revived. This kind of agriculture is a sound way to preserve biodiversity as well as minimize human impact so that areas that produce food can also be home to other living creatures.

Modern Sustainable Systems

There have been modern attempts to combine the sustainability, productivity and diversity of the older integrated systems with modern irrigation and management systems and very modern energy systems.

There have been varying degrees of success and failure in most modern sustainable endeavors but a broad philosophy of what sustainability is has emerged in the past few decades.

The philosophical underpinnings of sustainable agriculture include:

- a desire to integrate nature and agriculture
- synergistic integration between animal husbandry and agricultural activities
- strong cooperation between neighbors
- more independence for farm operators
- detachment from non-farm sectors
- use of local input
- decentralizing control of farm resources

- stimulation of the local economy.
- the use of renewable energy
- light-till or no-till strategies

Practices on the ground usually include:

- greater crop diversity
- crop rotation
- planting of soil building crops
- the planting of cover crops
- companion planting
- non-chemical weed control
- integrated pest management
- integration of crops and livestock
- intensively managed grazing to naturally recycle nutrients.
- avoidance of heavy equipment

Sustainable farming systems are very productive when one measures the output of high quality food. They are also resilient, flexible, and environmentally sound with few of the ecological problems of conventional farms.

Continuous monocrops only measure up when some creative accounting is done, such as not figuring in the cost of agriculturally based health problems and environmental degradation or ignoring farm subsidies and subsidized inputs such as gasoline.

Since there are as many variations on sustainable systems as there are sustainable farms, I will only mention two of the newest spins on sustainability: microgardening and permaculture.

Microgardening: The Underground Food Production Systems in the Soviet Union

At first I thought that microgardening should be classed as a form of urban gardening. But microgardeners are "creating an edible landscape" even if the "landscape" is the post-Cold War ruins of the worker's paradise.

Microgardens are sustained by human cleverness, recycled inputs and the need of the gardeners for wholesome food. By the square meter microgardens are seven times as productive as socialist-style industrial agriculture.

Microgardening is not a movement, it is a phenomena reminiscent of certain post-apocalyptic science fiction films in which small bands of people scavenge and survive in the ruins of modern civilization. Nevertheless it is a food-producing system, just as ignored by the authorities in former Soviet nations as subsistence agriculture is ignored by commercial and scientific circles.

Like subsistence agriculture microgardening feeds hundreds of thousands of hungry people who have been bypassed by the global economy and have no other way to get food.

Why Raise Rabbits on the Balcony?

In the Soviet Union the course of industrial agricultural took a very sinister term. In an eerie and slightly skewed imitation of the reductionist west the Soviet politicos decided that total standardization, collectivization and central planning would bring their nation the promised plenty of the worker's paradise.

When a cruel political philosophy was added to an already destructive agricultural system the results were disastrous. Forced collectivization led to the murder, exile and disgrace of millions of the Soviet Union's farmers. Those who remained were deprived of even the most basic incentives to work on collectivized lands and invested most of their energies in tiny private gardens.

Central planning created absurd situations such as nation wide sowing date for "new Soviet wheat" in the 1950s.

The wheat crop was to be sowed on the same day from Kaleningrad to Irkutsk, from the Black Sea to the Baltic.

In a nation with eight different climates and 12 time zones the results were predictably catastrophic.

Soviet tractors and combines were designed to be bigger than their Western counterparts. The result of this one-upmanship was the creation of machinery that was too big to be turned around in ordinary fields.

Ambitious cultivation plans and huge irrigation schemes increased the amount of land under the plow through the fifties and sixties but no money was invested in storage and transport. Consequently a third of the grain grown in the Soviet Union was never used by animal or man but rotted in the fields.

The environmental problems accompanying these policies were horrifying.

Agricultural production fell in all sectors.

Industrial agriculture plus central mismanagement had produced not false plenty but false scarcity in a nation with one third of all the world's best arable land. Standing in line for food and goods became a second career for the Soviet citizen.

This led to the rise of a powerful underground economy and a seething black market.

It also led to the truly astonishing microgardening systems which allowed Soviet citizens to grow cabbage, potatoes, tomatoes, herbs, beets, carrots, turnips, parsnips and tree fruits in extremely limited areas. The gardeners would use this produce to help them get through the winter or barter it for necessities unobtainable in government shops.

The ingenuity of the microgardener was truly fantastic. Not every Soviet citizen was lucky enough to have access to a private plot of land or a garden. Vacant lots, sheds, balconies, rooftops, were also pressed into service. Scraps of packing material and plastic film were turned into cloches, industrial rubbish became irrigation equipment. Small scale animal husbandry projects went on under beds and in attics.

The driving force behind this activity was sheer necessity.

Microgardeners were deprived of most normal inputs and sometimes persecuted by central authorities. In some areas the "underground" agricultural activity contributed more to the local economy than the government sanctioned farms.

The break-up of the Soviet Union has led to improvements. Some former Soviet republics are attempting to restore their agricultural production and offer aid and incentives to farmers.

Other republics and satellite states are still struggling with pollution, salinization, degraded farmlands and political instability.

The net result is that the former Soviet empire that should be feeding the world, cannot feed its own people and is the largest importer of grain, not for animal feed but for bread.

Applicability: Microgardening shows just how much food can be produced in minute areas when people are really hungry. But it should not have to be done this way. One way to improve the former Soviet Union's food supply would be to encourage the microgardeners with a few hundred dollars apiece of simple tools, seeds, and plastic water pipes. They would then be able to go out and grow the food their nations so desperately need. This would probably do more good than all the recent western aid to the disintegrating Soviet Union.

Permaculture

Photo by Elaine Solowey

Figure 9: Arid Permaculture

The word permaculture is a contraction of the words **perma**nent and agri**culture**. It is also a contraction of the words **perma**nent **culture,** based on the principal that cultures cannot survive for long without a sound agricultural base and a sustainable land ethic.

Permaculture attempts to draw together the wisdom of traditional farming systems, modern technical knowledge and the observation of natural systems into a cultivated ecology that is designed to produce food for animals and humans.

It is the philosophy of working with nature rather than against nature, of looking at plants and animals in all their functions rather than as a single product system, of using foresight and observation to save on labor and input.

Permaculture is not only an agricultural system but also an ethical one opposed to all forms of exploitation and waste. Care of the earth, care of people, and the life ethic, which recognizes the intrinsic worth of all living things, are core beliefs. Agricultural and energy systems are designed to conform to these beliefs. Consequently animals are kept humanely and provision is made for wildlife and companion creatures in agricultural areas.

The permaculture systems, once established are designed to function synergistically around the home and to raise most of what is needed within a short stroll from the farmers abode.

Some of the generally accepted tenants of permaculture are:

Think about long term consequences. Plan for sustainability

Use native species or natural species known to be beneficial

- Cultivate the smallest possible land area.
- Plan for energy efficient intensive systems

Be diverse, this provides stability

- Look at total yields of crops and animals in all their aspects.
- Regard energy saved as a yield
- Use solar wind and water energy or biological systems to conserve and generate energy
- Bring food growing back into cities and towns
- Assist self-reliance; promote community
- Reforest the earth and restore fertility to the soil
- Use everything at its optimum level

- See solutions: not problems
- Work where it makes a difference.

In essence permaculture is a revolution in lifestyle and mindset that attempts to change the way the farmer interacts with the world.

The laws and principals of permaculture design are drawn from ecology, energy conservation, environmental science and landscape design. They are briefly as follows:

Relative location: every element is placed in an assisting relationship.

Each element has multiple functions

Planning for the house and surrounding zones is energy efficient

Emphasis on the use of biological resources rather than non-renewables.

Energy recycling

Natural plant succession to establish fertile soils

Polyculture and integration

Use of natural patterns.

Permaculture design is based on seeing the connection between elements so it is possible to place an element where it functions most effectively. Dams and water tanks are located above the house so that gravity rather than a pump is used to direct the water's flow. Windbreaks around the home are placed so they deflect wind but do not deprive the home of exposure to the winter sun. The garden is placed between the house and the chicken pen so that garden refuse is collected on the way to the pen and chicken manure is easily shoveled into the garden.

Consequently a permacultural unit whether large or small is set up with much care and foresight and an eye for the human impact on an area.

The results of well run operations are impressive. Permaculture areas are beautiful and livable with food and useful material all around.

Applicability: A very adaptable system that shows great promise for bringing food production into the suburbs. Permaculture could be an excellent method of reclaiming land and strengthening rural communities.

Permaculture is not easily explained. For more information:

The Permaculture Way, Graham Bell, Thorsons, Hammersmith, London, 1992

Small Change, Marianne Frances, Center for Human Ecology, Edinburgh 1989

Natural Way of Farming, Mansanobu Fukuoka, Japan Publications, Tokyo, 1985

Designing and Maintaining Your Edible Landscape Naturally, Robert Kourik Metamorphic Press, Santa Rosa California, 1984

Permalculture One, Permaculture Two, Permaculture—A Designer's Manual, Bill Mollison, Tagari Press, Tyalgum, Australia

Another Kind of Garden, Ida and Jean Pain, Villecroze France, 1972

The Permaculture Plot, Simon Pratt, Permaculture British Isles, Buckfastleigh, England, 1991

The Common Factors of Edible Landscapes

Diversity, local knowledge and human skill are the key elements in all sustainable systems, all the edible landscapes.

Diversity Crop diversity gives a sustainable system a resilience unknown in modern farming. The Andean villagers who plant 30 varieties of potatoes know that some potatoes will do well, others not as well and some will do poorly. The next year depending on the climate, the numbers might be reversed so the villagers are as careful of the poorly producing potatoes as they are of the good producers. They consider all the kinds of potatoes they have vital resources.

The industrial farmer grows Russel Burbank potatoes over his entire farm. In optimal conditions he will have a marvelously bountiful crop of identical tubers. If anything goes wrong he might find himself out of business.

Local Knowledge Local knowledge is priceless in a sustainable system. The farmer knows which weeds choke out seedlings and which can shelter them, the farmer knows which wild plants will poison his livestock in the spring and which he can eat himself if he is hungry. Without this knowledge of local conditions farming would be impossible.

Industrial agriculture has wiped out the need for local knowledge, indeed it has wiped out the recognition that there is such a thing.

Standard plants are moved from the temperate zones to the tropics without adaptation. It is assumed that the same wheat that grows in Kansas must grow in India as well. Killing weeds is big business absorbing as much as a fourth of the budget for farm inputs. Very few conventional farmers actually know anything about the plants they are trying to eradicate.

The Human Element Edible landscapes are run on human stewardship and skill. Planting, sowing, reaping, building, directing, herding and harvesting; the need for the human hand, the human eye and mind is everywhere.

Industrial style agriculture is built around the million dollar machine and the multi-billion dollar transnational corporation, both as far removed from the process of making good food as it is possible to be.

Human skill is not wanted because skill has to be paid for. Human stewardship is considered a romantic and impractical notion that just gets in the way of making a profit.

The industrial system is currently the dominant one. But it will not endure. The price for this kind of "plenty" is too much for the world to pay.

It will not lead us to abundance.

Genetic engineering/genetic modification has been a great disappointment if considered solely as a means of improving existing crops or bringing new food sources into the world.

Touted as the ultimate feed-the-world strategy in the eighties and conjuring up visions of nitrogen fixing corn, drought tolerant wheat and a host of other made to order dream plants, genetic engineering has joined the ranks of controversial technologies that promise much and deliver little.

The primary reason for GM crops' uninspiring performance is the fact that GM strategies make food more *expensive*.

Anything that makes food more expensive to grow or acquire in the developing world is bad. Raising the price of the seeds, linking their use with chemicals that have to be purchased, bringing in costly technology for cultivation and harvest, all of these contribute to the problem of hunger. People are hungry not because there is not enough food in the world but because so many people lack the means to produce or buy it.

Modified seeds are costly and they are usually sold with a package of other agricultural inputs such as herbicides and fertilizers. Since GM seeds cannot be saved they must be purchased for each planting. It is obvious that the companies that own these seeds plan to collect on them wherever they are, forever, if possible.

An economic contract in which one party pays in perpetuity for the use of a vital resource is not unknown in human history.

It is called “serfdom” and is generally considered to be unfair.

Another reason that genetic engineering has not “delivered” is the fact that genetic engineering proved to be far more difficult than the pioneers of the science had assumed. Genetic engineers discovered that they could only move small snippets of genetic material, usually three to five genes at a time.

The agricultural characteristics that would have truly changed modern agriculture, the aptly named **“jackpot traits”** like salt tolerance and the ability to fix nitrogen are controlled by the complex interplay of many genes and cannot be readily transferred from plant to plant by cutting and pasting. The genetic foundation of these traits is not even fully understood.

Consequently sights were lowered and genetic engineering became the tool of the quick-fix, high profit school of agricultural science. Plants were designed to be resistant to Round-up so that a farmer might saturate his crop with herbicide, kill the weeds and still have something to sell. Genes from fireflies, moths, chickens, fish, worms, tobacco and bacteria were transferred to vegetables, corn, rape seeds and soybeans on the pretext of improving them.

Plans to truly improve crop plants were set aside and the feed-the-world rhetoric was forgotten for a decade and only resurrected when GM strategies were criticized by parties who feared that GM crops would endanger the food supply.

In the United States the FDA passed a ruling that in essence a potato is a potato even if it does have chicken genes. This notion is called "*substantial equivalence*" and means that GM foods are not different enough from the original to pose any health hazards. Some plants were tested. GM staples were fed for a few weeks to animals then declared safe for general use. This paved the way for the spread of genetically modified staples of all kinds, which the American public accepted without comment for the most part.

Then the US and the "Gene Giants" attempted to promote similar GM plants in Europe and ran into fierce opposition from farmers, consumers, environmentalists, conservationists and food retailers. The flash point of this acrimonious debate proved to be the galanthus nivalis agglutinin/snowdrop lectin (GNA) potato, a potato spliced with a snowdrop sequence to produce its own pesticide.

Is This Potato Really a Potato?

An experiment at Rowett Institute in Scotland indicated that the modified potato caused damage to the immune systems of rats. Not liking the news, Rowett Institute decided to shoot the messenger and dismissed Dr. Arpad Pusztai who had worked at the Institute for 36 years and was the author of over 200 scientific papers.

The experiment was declared "flawed."

It apparently never occurred to the proponents of GM crops and crop technology that what might be flawed is the world view behind the controversial technology.

Why GM crops Do Not Reduce Pesticide Use Other potatoes have been engineered to express Bt endotoxin and resist the depredations of the pesticide resistant Colorado potato beetle. But farmers who have tested plants have reported that resistant insects appear with astonishing rapidity. Bt resistant Colorado potato beetles, mosquitoes, budworms and bollworms crop up only a few seasons after the engineered crops were planted out.

This indicates that gene-spliced, built in pesticides are not any more durable that the sprayed on varieties. It does not matter to the bugs whether the pesticide is in the plant or applied. The insects soon become resistant.

To understand why all this time, effort and money is being spent on a strategy which has already shown itself to be a dead end one must return to examine the underlying assumptions.

Incidence of Resistance Proponents of Bt and GNA modified plants assumed that major resistance genes among insects must be rare. The evidence in the field undercuts this assumption. Resistance genes are probably more than 10 times as common as was formerly believed.

Dominance of Resistance Genes They also assumed that resistance genes must be recessive. There is evidence that resistance in insects ranges from recessive to incompletely dominant to dominant. This means, of course, that many species of insects will become completely resistant to Bt and GNA and some will do it **very** quickly.

Development of Resistance Parallels Exposure Pesticide secreting plants produce toxin in every cell during the entire growing season. This exposes insects of all kinds to the toxin as long as the plant lives. There is evidence that these toxins persist in the soil for up to a year as well. (Sears and Schaefema, 1998, pg. 2) Pest populations exposed continuously will develop resistance more quickly than pests that are exposed only during times of infestation. The fact that Bt toxin is present at all times in every part of the plant will soon render it useless as a pesticide.

If built in pesticides are matched so easily by the pest population then what possible benefit can be derived from the entire exercise?

Which brings us to the question: Is a poison secreting potato still a potato?

GM Plants are New Plants

A potato which contains a genetic sequence that causes it do something that no other potato can do is essentially a **new plant** no matter what it looks like.

New plants cannot be evaluated by assuming "substantial equivalence," but have to be tested for many qualities including long term toxicity. Certainly there seems to no point in developing a food crop that we cannot safely eat. Dr. Pusztai's experiment indicates that this might be the case with the modified potato.

It is not only the consumer who is at risk with these crops. There is some evidence that Bt secreted by modified crops persists in the soil and renders it infertile. Pollen from Bt crops definitely affects the fertility and life cycle of insects. The case of the monarch butterfly, which is killed by Bt corn pollen, is one example. Another example is the ladybug. Lifespans of ladybugs were reduced by 50 percent when they fed on aphids that were found on Bt potatoes. (Losey, Rayor, Carter, 1999, pg. 2)

Considering that crops are being engineered to carry genetic sequences from plants which are not edible, insects, mollusks, viruses, fish, birds and human beings one would think that long term studies and tests would have been carried out before the plants were released into the environment.

Why GM Plants May Be Incompatible with Human Health

There may be a simple and obvious reason why some GM plants are not fit to be food: many were designed after all, to poison insects. On the cellular level we are very like the rats and the butterflies. What is not good for them cannot be good for us.

Or there might be some deeper and more ominous cause for the damage done to the thymuses, kidneys and spleens of the experimental animals in Dr. Pusztai's study.

Lancet, the British medical journal decided to print Dr. Pusztai's report (Ewen, Pusztai 1999, pg. 1353). On February 12, 2000 the fact that Dr. Pusztai's findings had been verified by a panel of 20 scientists from around the world made headline news in Britain (Cummings and Liliston, 2000, pg. 3) and the potato, so to speak, was out of the bag.

Dr. Puzstai's experiment raised some troubling questions about more than the modified potato. When Dr. Pusztai fed conventional potatoes and snowdrop lectin to rats no damage occurred. But when Dr. Pustai fed the rats an equivalent amount of potatoes that were **gene-spliced** with the snowdrop lectin, the rats sustained immediate damage to their immune systems and many internal organs including their guts.

There may be something in the manner in which the snowdrop lectin was introduced into the potato genome that made it more toxic and dangerous.

Or there may be something seriously wrong with the entire process. Bt, spliced routinely into hundreds of thousands of acres of corn in the U.S. is considered a form of lectin. Without long term studies or any way of tracking the modified corn there will be no way to tell if modified corn is hazardous to the health

California Mosaic Virus may be the culprit (CaMV). It is commonly used to vector genetic material in the gene splicing process. There have been no studies relating to the ingestion of material modified by CaMV. Automatically assuming that these plants are not latently toxic seems very irresponsible.

Antibiotic resistance marker genes, used to monitor genetic sequence transfers, are of special concern to many medical doctors.

The British Medical Association which represents 115,000 doctors, requested a moratorium in a published statement on May 17,1999, (Frith and Murphey, 1999, pg. 6) on the planting of GM crops because of the risk that the marker genes may encourage the development of drug resistant bacteria or be transferred into intestinal microflora. The doctors also called for a ban on releasing genetically modified organisms (GMOs) into the environment and demanded a review of the World Trade Agreement to ensure that human health and safety take precedence over interests of global trade in the sale of foodstuffs and seeds.

Why GM Plants Do Not Increase Yield

GM crops often yield less, not more than ordinary crop plants, because the plants that are selected are not chosen for high yields but for the ability to accept alien sequences. This quality is called "transgene tolerance" and explains the thousands of failures in the initial stages of modification which makes genetic engineering so costly. Most organisms are not transgene tolerant.

There is also a metabolic cost to the plant which must express herbicide resistance or produce Bt toxin. The resource debt the modified plant incurs to do this tends to depress the harvest index.

This lag in productivity is called **yield drag** and was documented in modified Roundup Ready soybeans in the Ontario Soybean Variety Trails in 1999. (Oplinger et al, 1999, pg. 1)

A combination of yield drag and Bt resistant bollworms caused cotton farmers from Georgia to Texas to lose money in 1998 and 1999 since the modified plants did not produce as well as the unmodified plants, were not impervious to bollworms as promised and produced a commodity which many people would not buy. (Students Against Genetic Engineering, 1999, pg. 2)

It is difficult to see how expensive plants, some of which may not be usable, that produce less than ordinary unmodified ones are going to feed the world.

The Whitebread Syndrome

Perfectly good wheat is stripped and milled and bleached to become white flour. The "millfeeds" which contain most of the food value are sold separately; a few cheap synthetic vitamins are put back into the almost worthless flour and then the flour is considered "enriched."

"Enriching" something slightly after impoverishing it totally does not seem to me an honest practice.

The newest spin on the "Whitebread Syndrome" is the "golden rice" project.

On the surface it looks like an admirable project with a humanitarian aim, to save people who have little to eat but rice from blindness caused by Vitamin A deficiency including one million children. (Gillis, 2000, pg. 4)

Unfortunately the project is designed in a manner which is sure to make it a disaster.

Why does the target population have only rice to eat?

Some of the people in question are landless due to the collapse of traditional agriculture and can afford no other food as they scratch for a living in the shantytowns of Asia. Other people are still living on the land but have been forced to accept a chemical version of paddy culture which has eliminated the fish, algae, and greens that were once raised with the rice. The algae/azolla culture provided the rice with nutrients but the fish and greens were consumed by the farmers themselves. Modern paddy culture may squeeze a few more kilograms of rice out of each paddy, but it has eliminated the high protein food and the high vitamin food which kept the farmers healthy.

What is wrong with the solution? Exactly the same thing that is wrong with "enriched" bread. Something important and irreplaceable has been taken away and now the people who are suffering are offered an inadequate and inferior replacement.

This is not a workable solution for the following reasons: First, *the target population is insufficiently fed.* Lack of Vitamin A is only the most obvious of the deficiencies. To improve the situation of the target population the other essential foodstuffs must be supplied as well. Will the "golden rice" also be engineered to supply Vitamin C, Vitamin D and extra protein?

Second, rice is such an important food to the target population that tampering with the taste, smell, texture or cooking qualities might be catastrophic. There is no evidence that the target population will agree to eat the engineered rice. For the West to then tell them they had better learn to like it would surely be adding grave insult to injury.

Furthermore this project accepts, which I do not, that there will be populations in the 21st century that will just have to get by on rice alone.

I also must ask if engineering a staple that feeds half the world may not have unforeseen effects when the "improved" version is planted out. We are, after all, not speaking of a gene pool but a mighty river of plants and grain which flows from the past into the future. Polluting this river with alien genetic sequences may have serious consequences for all mankind.

Lastly, I must ask if this is a sensible way to spend 30 billion dollars which is what the project will apparently cost even though many biotech firms have agreed to "donate" the technology.

Surely a much less expensive, invasive, meddlesome and patronizing way can be found to deal with this hungry population.

For example, a modest program to encourage the cultivation of all those wonderful A-rich Asian greens that were once available or a program to help them cultivate spinach, carrots, sweet potatoes and spirulina.

Giving them microcredit loans of a few hundred dollars apiece to help lift them out of their desperate poverty would also be possible with this sum of money.

I find it hard to believe that the biotech corporations understand the target population's needs better than they do.

Has anyone asked **them** if **they** want "golden rice"?

Can GMO's created for animal husbandry be kept out of our food?

StarLink corn was created by Aventis CropScience. StarLink corn carries a genetically modified protein which protects it from corn borers. Unfortunately the protein, Cry9C, is not broken down by gastric juices and can trigger allergic reactions. In the fall of the year 2000 StarLink was found to be mixed into the human food supply even though it had not yet been approved for human consumption. A massive and costly recall of a dozen food products followed including tacos, corn dogs, beer and corn snacks. After reviewing evidence from many cases of allergic reactions an EPA expert panel concluded that there was a medium likelihood that StarLink could indeed trigger such reactions. The panel then decided that since there was such a limited amount in the food supply that there would not be very many cases. Then the panel recommended that the EPA not act on Aventis's requests to approve the food for human consumption until a test was created and used to evaluate the reports of allergic reactions. (*Kaufman, International Herald Tribune, March 30, 2001*)

This is an unsatisfactory situation. A medium likelihood of allergic reactions should insure that StarLink is **never** approved for human consumption. The probability of such reactions will only remain low if the booby trapped corn remains a tiny fraction among the unmodified corn in the food supply. If the percentage of Starlink in the food supply increases the probability that some people will be made ill and die increases as well. There will be no way to avoid the corn since most GMO's are not labeled but just jammed willy-nilly in with unmodified foodstuffs.

The StarLink controversy not only demonstrates the difficulty of keeping GMO's out of our food when the law is on the side of the consumers **but totally discredits the notion of substantial equivalence**.

StarLink looks like ordinary corn but it is not and it is different enough from ordinary corn to kill or damage an unknown percentage of the people who eat it.

Since the protein cannot be digested by the animals it is fed to—is also possible that the Cry9C protein will be just as allergic second-hand.

In a sane economic system there would be **no room** for Starlink in the food chain.

So what are GM crops good for? Some GM crops have been successfully designed to enhance quality traits. For example there are GM tomatoes with more solids, more sugar or a lower freezing point. But these traits could have been bred into tomatoes without using genetic material from unrelated organisms by cross breeding with existing tomato plants which carry those traits. Most transgene traits are "input" traits such as Bt or GNA that are supposed to save the farmer money on pesticides. Other desired "input traits" are tolerances to specific herbicides.

If a person doesn't want to ingest poison, **inside** or **outside** of the crops then current GM plants aren't good for much.

Do the modest benefits GM crops can deliver justify the terrible risks? This question has to be asked because the existence of GM plants expose us, the crop plants that we depend on and the eco-systems of the world to many separate categories of risk.

Risk to the Consumer The risks to the consumer are obvious. If the Bt potato has negative effects on the immune systems of rats then it is almost certain that it is not fit for human consumption. Human beings are not the strongest creatures in the animal kingdom but some of the most fragile. For this reason alone the "flawed" research from Rowett Institute must be taken seriously and the experiment repeated until it is known beyond a shadow of a doubt whether GM crops are safe when used as food.

It is not the responsibility of the consumer and the environmental protection agencies to prove that the GM crops are dangerous.

It is the responsibility of the proponents of GM strategies to show that these crops are fit to be food.

Every strain of GM crops must be tested in this manner because a familiar plant containing the genes from an unrelated organism is not what it appears to be. A GM plant is in essence a new plant and should be treated as such regarding toxicity and nutritive value.

Direct Environmental Risks How does a GM plant affect the living things which come in direct contact with it? GM plants may be more competitive than indigenous plants in an unmanaged ecosystem, "get loose" and become invasive. This has already happened with dozens of imported exotics. Adding GMOs to chaos cannot make things better.

Herbicide resistant GM invasives in wild ecosystems may have truly nightmarish effects, wiping out indigenous species and fundamentally changing the populations of insects and animals.

It is all very well to boast about the benefits of an insulin producing sugar beet for humanity. But what happens to the bees, soil organisms, birds, and animals that come in contact with such a plant?

We already know the pollen of the Bt plants is toxic to some organisms and stunts the growth of others. What will be the effect of GM plants which are designed to produce pharmaceuticals?

It is possible that GM nutraceutical crops cannot be grown safely in the field for this reason.

Again the only way to determine what is safe is by extensive testing of each and every GM variety in *closed* formats, letting them interact with insects, birds, animals and other plants without releasing them into the general environment.

Risks of Field Testing The GM plant may become invasive in a new environment, seeded by the plants involved in the test. The small scale of the test patch will not tell teach us how the GM plants will affect their wild crop relatives when planted on a commercial scale. In general, small scale testing risks too much and teaches us too little.

Specific Risks of Commercial Planting Commercial scale planting exponentially increases the opportunity for gene transfer while making the careful monitoring for "escapes" far less effective.

Commercial sized crops will be also exposed to a wide variety of weather events which may spread GM material far beyond the confines of cultivated lands.

Therefore commercial planting should not have even been contemplated until careful testing gave us some idea of what would happen in the interaction between GM plants and the natural world.

Virus Resistant Crops

Many biotech labs are working to engineer plants resistant to the damaging effects of plant viruses. To do this GM engineers splice viral genes into the plant's genome. The resulting GM plant is resistant to infection by the virus from which the viral gene is taken.

At first glance this seems like a good idea, a sort of anti-virus vaccination. But the mechanism by which immunity is conferred is not fully understood. The strategy relies on the unexplained capacity of the viral genes or gene products to confer a protective effect once the plant begins to synthesize them. Usually virus coat proteins are used in the engineering but scientists have also used small pieces of ribonucleic acid (RNA).

But plants can be infected with more than one virus at a time. In some cases the coat protein of one virus can enclose the genome of another, a rare phenomenon called transencapsidation. This creates a new virus with the genome of one disease but the coat protein of another.

RNA conferred protection can also be problematic as the RNA can combine with other virus associated molecules and create a virus which is resisted by the host plant but can be passed to other plants.

As a result new viral strains may arise or the host ranges of existing viruses may be broadened. (Rissler and Mellon, 1996, pg. 60)

Risks to Crop Plants and Wild Relatives The presence of GM crops threatens the genetic purity of unmodified crops and wild crop relatives. So far no guidelines have been established regarding the distance that must be maintained between modified and unmodified crops to prevent gene transfer. There are no guidelines relating to buffer zones or buffer crops. No one has any idea how quickly or how far GM sequences can be spread. Consequently planting a GM crop is like sowing the wind. Lest the old axiom from the Bible prove true and we reap the whirlwind all these possibilities have to be investigated thoroughly. (The Bible, Hosea, 8:7)

Some possible consequences:

GM crops could compromise unmodified crops in the area

Monsanto has demanded substantial sums of money from farmers who have never planted their GM seeds. Many of these farmers claim that they have never bought GM seeds and have no intention of doing so, but that their crops have been pollinated by GM crops in the area of their farms.

In several cases this has proved to be true yet the farmers had been forced to pay royalties for GM seeds they didn't want or purchase. Other farmers have refused and ended up involved in costly lawsuits as the biotech companies try to protect their investments. (Leonard, 2000)

Anti-GM lawsuits have also been filed by organizations and individuals, many who claim that the FDA has violated their right to avoid eating GM foods by stonewalling on the labeling of GM crops and mixing GM farm products with unmodified ones. (Coale 2000; Brasher, 1999)

GM Plants Which Have Escaped.

In 1994 Monsanto, one of the world's largest agrochemical companies, perfected a genetically altered variety of rape seed called Round Up Ready. This variety could be sprayed with Round Up herbicide and sustain no ill effects. Ordinary rape seed plants are killed by an application of Round Up.

Monsanto arranged for a number of farmers in Sasskatoon, Canada to grow the new crop. Each farmer paid a fee and signed a contract agreeing not to save any seed for future planting and not to give seeds to anyone else.

In 1998 Monsanto discovered that farmer Percy Schmeiser was growing the GM rape seed. Schmeiser had been growing rape seed for years and had always saved some of his seed for next year's crop.

Monsanto demanded US $29,000 in royalties,

Schmeiser claims he never planted *Round Up Ready* rape seed. The strain in his fields was apparently a result of his rape seed plants being pollinated by the GM plants in nearby fields.

Monsanto has tried to prove that Schmeiser obtained the seed from another source.

Schmeiser says though the GM strain does save labor it does not give as good a yield as his own selected strain. *By rights I should sue Monsanto for allowing their plants to invade my fields* (Weiss, 1999).

The case recessed until the autumn of 2000. The amount of money Monsanto was demanding was in the hundreds of thousands of dollars. Schmeiser refused to settle out of court because he claimed he never planted any of Monsanto's seeds. (White and Desmonie, 1998)

Ultimately the judge came down on Monsanto's side and Mr. Schmeiser was ordered to pay for the use of genetically modified plants he never bought or wanted, held to a contract he never signed.

This does not bode well for any farmers downwind of fields where GMO's are planted.

This incident also calls into question the claims that GM plants will not "escape" and poses the quandary of ownership of these plants once they have spread from their point of introduction.

Wild Relatives Pick up GM Sequences That transfer of GM traits is possible has been confirmed by the appearance of herbicide resistant weeds near GM rape seed fields (Western Producer, 2000). Other GM traits may prove even more problematic.

The wild relatives of GM crops may become invasive This *is* a real possibility if a transferred GM sequence confers an advantage to a wild plant, making it more competitive or accelerating its growth. (Rissler and Mellon, 1996, pg. 49)

GM sequences can flow into the centers of plant biodiversity If GM sequences flow from wild relatives into the centers of crop diversity the very genetic reservoirs we need to develop new strains of crops will be compromised. (Rissler and Mellon, 1996, pg. 67) We must ask ourselves if we have the right to deprive our descendants of the tools they will need to cope with future challenges in agriculture. This is just what we will be doing if we allow the centers of crop diversity to be contaminated by GM genes.

Cascading and Cumulative Risks Dichloro-diphenyl-trichloroethane (DDT) the “Miracle Chemical” of the 1950s cascaded through the food chain and the world’s ecosystems so quickly and catastrophically that the price paid in disease, extinction and suffering is still being totaled up today. (Null, 1990, pg. 111)

DDT poisoned insects, blighted the lives of insect-eating creatures, accumulated in the tissues of predators and ended up in the breast milk of every woman on earth. Traces of DDT are found in the deep waters of the ocean, on Mt. Everest and in the Arctic circle. (Null, 1990, pg. 109-110) As pervasive as this substance is, it is an inert material, it cannot reproduce itself nor mutate nor spread itself on the wind.

GM strategies change the stuff of life itself. GM plants seed themselves, mutate and transfer genes to related plants. Even the proponents of GM strategies admit that they cannot foresee all the problems that might arise as the GM sequences accumulate in the lines of crop plants or the effect GM plants have on other organisms as they ripple through the web of life. (Independent U.K, 2000; Associated Press, 2000; CNN, 2000; BBC News, 1999)

Robert Shapiro, the chief executive officer of Monsanto said on October 28, 1998:

We don't seek controversy, but obviously it has been thrust upon us. It is a direct consequence of a role we have chosen. And it is a role which we can only blame ourselves for...we realize that with any new and powerful technology with unknown, and to some degree unknowable—by definition—effects, then there necessarily will be an appropriate level at least, and maybe even more than that, of public debate and public interest. (Monsanto Monitor, 1999, pg. 7)

Epistasis, Pleiotrophy and Transposons: The Wild Cards of Genetic Engineering

The term "pleiotrophy" means that a single gene product can affect more than one trait in an organism. The outcome cannot be reliably predicted but is only discovered when the genes interact in a plant or animal. In the 1970s thousands of acres of corn were devastated by corn blight because of an unexpected pleiotrophic effect. T-cytoplasm, a male sterility gene, also suppressed blight resistance in corn allowing blight to sweep through a vulnerable crop population. At that time three-fourths of the corn in the US carried the male sterility gene. The crop loss in the Corn Belt and the southern states was severe and could not have been predicted. (Raeburn, 1995, pp 132-136)

Epistatic effects refer to the ability of one gene to influence another gene which is not an allele of the first. An example is the production of purple flowers when two white sweet pea varieties are crossed. Two separate genes interact to produce the color; neither gene can produce the color alone.

Transposons are recently documented "jumping genes," genetic sequences which are spread by horizontal gene flow or non-sexual transmission between adult organisms. While horizontal gene flow between plants is relatively uncommon, gene flow between plants and microorganisms has been observed sufficiently to raise the alarm about interactions between GM plant material and the microorganisms in the digestive systems of mammals (Rissler and Mellon, 1996, pp. 68-59).

Pleiotrophic, epistatic and transposonic effects will occur with GM plants and will be detectable only in the living organisms which are affected. The difficul-

ties in predicting these effects makes risk analysis for GM crops even more complex.

There is something disturbing about the eagerness of the proponents of GM crops to force the modified plants on the agricultural world and the consuming public without proper controls and studies. The fact that most corporations do not want these products labeled is even more suspicious.

The entire GM program seems to be an effort not only to buy, sell and own the basic stuff of life but to limit our choices and **force** us to eat these things.

How will GM Strategies Affect the Lives that Revolve around Farming Activity?

The twentieth century has not been a good time for the farming population or the farming environment. Modern agriculture has been ruinous for both. The environmental costs are well known. Farming populations have greatly decreased in the developing world and plummeted to unimaginable lows in developed nations like the United States.

Will GM crops accelerate the trends which concentrate more power, land and resources in fewer and fewer hands? Will GM crops cause more environmental disruption? Will GM strategies exacerbate the tensions between the nations which produce raw materials and the nations which purchase them?

Unfortunately the answer to all of these questions seems to be yes.

Genetic Imperialism One of the unfortunate side effects of GM activity is the escalation of the bitter war of words and deeds between the developed world and the farmers of developing nations.

GM crops are seen as by many as "genetic imperialism."

The farmers of the developing nations feel with some justification that their countries have been turned by old style military imperialism into mere depots of raw materials, their own crops and agricultural systems having been replaced over the last few centuries by the kind of crops which brought wealth to the developed world but only debt and perpetual food insecurity to them.

The gene pirates of the 20th century have added fuel to the fire, "discovering" plants, medicines and genetic sequences in the developing world and then patenting them in a manner which gave no credit to the indigenous people who first used and discovered them. Sometimes the patents prohibit the indigenous people from making further use of their own plants.

Billion Dollar Trees A prime example of this is the neem tree. Several of the neem's chemical compounds have been patented by American companies who have profited greatly, making soap, shampoo toothpaste and pharmaceuticals from the neem tree's seeds, bark and leaves. This interferes with the development of neem products in India where the plant originated and has been used for thousands of years. (FRLHT, 1995, pg. 13; patent on "neem," 1999)

Fortunately the patent on the neem was rescinded in the spring of 2001 and now this marvelous plant can be used and developed without interference.

The Strip-Mining of Genetic Wealth Landraces and heirloom varieties of crop plants have been used to improve or save mainstream crops, making millions of dollars for seed companies which rarely acknowledge and almost never compensate the indigenous farmers who originally used the plants.

As this is happening mainstream agriculture is eliminating the very landraces, relic crops and heirloom varieties it depends on for future development, putting the farmers who feed themselves with these plants out of business.

Crop biodiversity is not a concern of the biotech firms. Their already impoverished library of trendy, profit-now crops often eliminate strains of plants more valuable to farming communities; the same way that cash crops pushed out the plants that once nourished and clothed the people of the developing world.

Suicide Seeds Even more feared and resented than gene piracy are the GM sequences known as the "terminator" genes. These "property protection" strategies are composed of genetic sequences which produce toxins under certain conditions. These toxins render the seeds of a plant sterile ensuring that the farmer who wishes to grow this crop has to buy seeds from the company every year. (Dougherty, 1999)

Farmers in the developing world are seed-savers by necessity. Transfer of the terminator gene could compromise whole lines of food crops rendering them useless in the second generation. The Monsanto Terminator project was apparently discontinued in 1999 according to company spokesmen but other seed companies are still working on their own versions of "property protection" sequences which are designed to work differently. (BBC, 1999; New York Times, 1999)

The new genetic sequences produce a toxin in the second generation which makes the plant useless for food. The possibility that these genes will transfer themselves into the staple crops of the developing world, effectively booby-trapping them, is indeed a horrifying one.

Plants That Kill Their Own Offspring.

Technology Protection System (TPS) is a clever three gene system producing plants which kill their own offspring.

This is the most hated and feared of the GM sequences and has caused farmers in India to go on a rampage and burn crops suspected of carrying these genes. (Weiss, 1999)

Dubbed the "terminator" sequence it forces plants to produce a toxin that is fatal to the plant's own seeds.

The poisoning occurs late enough in the year to allow the seed to retain its value as food but the seeds do not sprout and farmers cannot save seed from year to year.

Since the seed companies need to produce seeds

from these plants, the suicide genes are under control of another genetic mechanism which suppresses the toxin indefinitely. In the suppressed state the plants produce fertile seeds. Just before the seeds are sold they are sprayed with a chemical inducer which overcomes the suppressor and activates the dormant seed killing gene. The seeds grow into plants which produce toxins which render their own seeds useless.

The "terminator" gene sequence is seen as a threat to the one billion subsistence farmers of the world who will never be able to afford to buy seed each year.

The Consultative Group on International Agricultural Research, (CGIAR) which develops new seed varieties for the developing nations has banned the technology in its seeds and research programs.

Once you've got farmers hooked on it and they've lost their traditional varieties, it is very hard to go back again, said Geoffrey Hawtin director general of CGIAR's International Plant Genetic Resources Institute in Rome. *Then these companies will be sitting pretty on a captive market.*

Other groups have asked for a global ban. Mark Ritchie, president of the Institute for Agriculture and Trade Policy in Minneapolis said, *It is a threat globally to food security which is a basic human right.* (Cohen, 1999)

Anti-GM critics worry that pollen from TPS plants could fertilize crop relatives causing epidemics of crop sterility, epistatic and pleiotrophic effects and a flow of dormant seed killing genes into subsistence crops.

Pro-GM scientists have declared that TPS systems can be a possible solution to hunger in the developing nations because if seed companies are assured of a market they will sell their best varieties to the farmers of nations where patent protection laws are weak.

Harry Collins, vice-president of Delta & Pine Land, the first company to use the TPS gene sequences said: *This will give the subsistence farmer a superior crop. He will get a crop that allows him not only to subsist but to be productive*. (Cohen, 1999)

Pro-GM scientists discount the fears of other groups even though they admit that they cannot predict how the "terminator" genes will interact with wild and weedy relatives of crop plants.

Accusations and Boycotts GM strategies have also caused hard feelings between the US and the European Union (EU).

US beef has been banned in the EU because of its high hormone content. The EU also objects to GM corn and soybeans and has demanded strict labeling of US products to insure that no GM materials are included covertly in processed products.

Rather than seeing this as a valid health concern the US has petulantly put 100 percent tariffs on several European foods including Roquefort cheese and foie gras. The EU's legitimate doubts have been interpreted as anti-Americanism. Senator Richard G. Lugar of Indiana who headed the Senate agricultural committee has gone as far as calling it "collective madness." US government sources seem determined to blur the line between GM crops and the products of conventional breeding processes. The above named senator from Indiana claims that corn and soybeans have been modified on his farm since the 1930s even though the GM technologies did not become possible until 50 years later. (Cohen, 1999) This allows GM proponents to claim that altered staples have a long and respectable history and European fears and feelings to be written off as irrational and anti-progressive.

Foul Food But Europeans are more discerning and sophisticated about their food than Americans and have long regional traditions in cultivation, animal husbandry food processing and cooking. The EU has also suffered from one food contamination scandal after another from salmonella in eggs and mad cow disease in England to poisoned Coca-Cola and dioxin-laced chicken in Belgium. They are in no mood to accept the US's truth stretching exercises having suffered for so many years from their own food follies.

So far the more militant protesters have limited themselves to torching and trampling GM test crops and trashing McDonald's restaurants. Some anti-GM activists have ended up in jail because they do not believe that GM crops and unmodified crops can coexist. Despite the obvious silliness of some of the forms of protest (and many of the protesters) there is a certain amount of truth to this notion.

Jose Bove of Montpelier who claims he is leading the "foul food" revolt turned himself in to the French police after ransacking and destroying a new MacDonald's on August 12, 1999. On September 28th 2000 he was sentenced to serve six months in prison for his offense.

Patrice Videau, the secretary of the Peasant Confederation, a growing farmers movement founded in 1987, has denied being anti-American. *What we reject is the idea that the power of the marketplace becomes the dominant force in all societies and that multinationals like McDonald's or Monsanto come to impose the food we eat and the seeds we plant.* (Cohen, 1999)

In Seattle the World Trade Organization (WTO) talks were disturbed by demonstrations, riots, sit-ins, and protests of a less conventional nature; many of them specifically against GM crops.

Indian activist Vandana Shiva has claimed that GM strategies are no less than a corporate hijacking of food and agriculture which is depriving millions of people of their livelihoods. Corporate monocultures create hunger rather than satisfying it by displacing integrated crops and eliminating biodiversity. The WTO with its secret decisions, aggressive export policies and obvious pro-corporate bias signifies for her the emergence of food totalitarianism in which a handful of powerful bodies try to control the entire food chain and decide literally who will live and who will die. (Shiva, 1999, pg. 6-17)

William Pfaff defines globalization in an International Herald Tribune editorial: *Globalization—the aggressive program for the imposition of Western norms of economic management, economic deregulation and market opening, and facilitating takeovers of indigenous industries and agriculture—was launched by the Clinton administration during its first term.* (Pfaff, 2000, pg. 6)

An integral part of the supposed takeover plan was an equally aggressive drive to make the world safe for GM technology. Not surprisingly the biggest advocates of GM strategies were usually ardent supporters of the globalization drive.

The reluctance of GM proponents to accept labeling laws, allow studies and wait until there are clear answers to questions of public safety has created an atmosphere of distrust and suspicion.

Many people worldwide feel that GM crops are literally being "forced down their throats" with little thought for the consequences.

With billions of dollars at stake on the pro-GM side and the public interest at stake on other, no one is in the mood for the sort of cooperative research that could map out the true risks of GM strategies and come up with a workable set of rules.

Nitrogen-fixing corn will most probably remain a dream, salt tolerant wheat and other plants that could truly revolutionize agriculture will probably never be created.

GM strategies have become the property of a few large companies whose prime interest is maximizing profit and are willing to risk environmental damage and public health and safety to do it.

It is difficult to abandon a technology that has so much promise. But this technology is obviously in the wrong hands and is being used to modify the wrong plants for the wrong reasons.

It is not possible to "engineer" the path to abundance.

Decrease the Risks-Possible Rules and Regulations for GM

Do not transfer sequences from inedible plants into edible ones An ancient Talmudic tractate describes the hybridization of a dog with a wolf as unfair to the animal, and the environment because the results of the mating would be a *wolfish dog*, a danger to humans and flocks or a *doggish wolf*, unable to hold its place in the pack. A mixing of this nature is called *kil-iy-im*—a *forbidden mixture*. Produce which looks like a familiar food but it radically different from the original and may not be safe to eat is surely as dangerous as the wolfish dog. Why create such chimeras?

Make GM plants sterile Most GM organisms are not going to be used in breeding—they are meant to be planted out, to be used and consumed. Since one of the main categories of risk is in the accidental transfer of engineered sequences to wild plants and crop relatives it surely makes sense to engineer sterility factors as well. This is doubly true of GM animals.

Do not engineer human genes into GM organisms We do not need to jam human genetic sequences into plants and animals just to show how clever we are. There are better ways of producing vaccine and heart valves then mixing our own genes with the genes of bananas and pigs.

Do the pharming in closed formats Most of us would prefer not to be vaccinated, inoculated, immunized or medicated when we eat an apple. If medicinal substances are produced by crop plants then the modified crops must be grown in closed formats and made sterile and all produce from them registered and labeled.

Do not insert the genes of insects and animals into GM crops Many of us have religious and ethical limitations on what we are willing to eat. Since we are the ones who are paying we would like our corn unseasoned with firefly genes and our fish untangled from our chips, thank you very much. Limit the gene swapping to the edible plant kingdom when working with edible plants and label the produce so the people who have sensitivities and allergies can make informed decisions.

Do not booby trap the future Ban Terminator genes and all the nasty genetic surprises that express themselves in future generations of plants. We have no idea what plants and crops will be needed in the future. Surely we should leave the playing field as level as possible for our descendants and not littered with genetic mistakes, traps and minefields.

Throw "substantial equivalence" out the window Potatoes with fish genes or snowdrop crocus genes are not the same as ordinary potatoes. They have been changed, sometimes invisibly into tubers from a new plant. Stop pretending that these changes are not important to the consumer.

Get the ownership issues right So you have made a new kind of corn or soybeans. And you own it. Fine. Just remember that you own it if it transfers something unpleasant into a neighbors crop, you own it if something in it poisons the ground, you own it if someone dies from an allergic reaction because you didn't tell them you have spliced in peanut genes. You own it and you own the potential risks, the potential damages You do not own it just when it is time to collect money...the cost of clean up is also yours.

Don't dare say to us when things go wrong that it is "just corn."

XIII. Single-Celled Wonders: The Promise of Edible Algae

Sometime around 7,000 BCE the Sumerians learned to brew beer.

Even before that nomadic people were practicing the most basic forms of cheese making.

In 4000 BCE Egyptians began to leaven bread with yeast.

The Israelites kneaded sourdough loaves just before the Exodus. They did not have time to let the bread rise and baked the still unleavened bread to take with them in their flight.

It is apparent from these four examples that humans had been using very small organisms to make various foods for thousands of years.

The yeasts used in bread and beer, the molds used in cheese making are just two of a host of organisms that inhabit the air, the soil, the water around us.

Soil symbionts in fertile earth, minute fungi which encourage plant growth, nitrogen fixing bacteria that inhabit the root nodes of legumes, the benign flora of our own digestive tracks—there is an immense invisible kingdom to be explored all around us. And we are just now beginning to understand how important these tiny organisms are to the ecology of the world.

Can some of these tiny creatures not only help us produce food but be food for a hungry world?

Letting Bacteria do the Work for Us The ultimate dream of the biotech industry is producing everything from artificial seeds to basic foodstuff in vats. This, of course, would do away the need for most forms of agriculture. Not only farmers would be eliminated. We would not need the sun, the rain and the apple seed, either.

A "bug" would be designed to do the work for us. Biotechnology would feed the world.

Like most high-tech shortcuts to abundance this has proved to be more difficult than was initially expected. Producing vast quantities of useful material by use of modified micro-organisms is still expensive and hedged about with formidable technical difficulties Single Cell Protein technology which was supposed to make the former Soviet Union self-sufficient in cattle feed by 1990 has done no such thing even a decade and a half later. (Hobbelink, 1991, pg. 94) Artificial seeds produced by fermentation are too expensive to be used in mainstream agriculture. Amino acids such as tryptophan have been produced by novel methods but in one well publicized case the end product was contaminated by an unknown compound that caused death or injury in those who ingested it.

Algae as Food for Man and Animals There are micro-organisms that can already be used for food safely, easily, and in a ecologically sound manner.

They do not have to be engineered or modified to be of benefit to humanity any more than the wheel has to be reinvented. They already exist in useful, elegant and safe forms among the myriad strains of algae.

Can they be produced in sufficient quantity and quality to make a difference?

Mass Cultivation's Promise and Problems Certain algae are suitable for mass cultivation (colorplate 23 on page 268, colorplate 25 on page 269). But there are drawbacks which limit the expansion of such endeavors. For one, because mass cultivation requires investment in a complex system of raceways, pipes, tubes, ponds and drying equipment there are only a few products which can be extracted from algae which are valuable enough to justify cultivation.

Photo by Elaine Solowey

Figure 10: Spirulina Raceway

A lack of know-how and experience has also limited mass cultivation. The technologies and tools are still being developed

There is also the problem of the purity of the algal culture. Just as a cornfield provides a comfortable environment for weeds, algal culture media provides an opportunity for all kinds of single-celled "freeloaders" some of which make the culture unusable.

Photo by Kuthi Kubenstein

Figure 11: Propagation

The closed algal bioreactor which seems to be a solution to this problem has a set of problems all its own, including the tendency to overheat.

Still, despite these difficulties, three species of microalgae are already being cultivated on a large scale.

Chlorella: The Green Pioneer

Early trails of algal biomass production were done in the 1950s at the Tokuga Institute of Biology and at Carnegie Institute in California. These systems all concentrated on the fresh water algae chlorella. The mass cultivation system pioneered in those experiments were later used to cultivate many strains of algae with many potential applications including food production, feed, extractable chemicals, aquaculture, nutriceuticals and wastewater treatment.

The first photo bioreactors and tube gardens were seeded with chlorella and the first attempts at raceway and pond cultivation involved chlorella as well. Chlorella was considered at first the most likely of the nine promising freshwater algae that had been used as human food. But chlorella production never made the transition from applied research to commercial mass cultivation as a food plant.

When grown outdoors chlorella cultures quickly became contaminated with all sorts of other microorganisms: spores, zooplankton which eat algae, single-celled animals and other strains of algae, some of which were toxic. It seemed impossible to produce a culture pure enough to use as food. Chlorella also has relatively strong cell walls which makes digestion by humans difficult. While cultivation and research of chlorella has continued for non-food purposes attention soon turned to another species of freshwater algae, the blue green cynaobacteria.

Current Use: Chlorella is cultivated for extraction of beta carotene, chlorophyll and xanthophylls. It is used in freshwater aquaculture and wastewater treatment.

Potential Use: Chlorella is a potentially valuable element in "biosphere" experiments because of its place in the freshwater food chain. It can be successfully cultivated in closed systems.

Comment: Chlorella is probably not suitable for mass cultivation as food for humans but it is a valuable food plant for freshwater fish and may be valuable in the future in ecological restoration projects.

Blue Green Algae

These organisms have weak digestible barriers between cells composed not of cellulose but of soft mucopolysaccharides. Mass cultivation became feasible because of the affinity of a certain type of cyanobacteria for highly alkaline water.

Spirulina: An Ancient Food from Alkaline Waters

Spirulina has a venerable history as a life form. It is a descendent of the first blue-green algae which appeared 3.6 billion years ago on a raw young world covered with water and saturated with carbon dioxide. It is assumed that cyanobacteria and other algal forms played a great role in the formation of an oxygenated atmosphere. The first photo-synthesizing cynaobacteria are abundantly represented in the fossil record as filaments of single cells stacked end to end, looking very much like spirulina does today.

Spirulina also has a respectable history as a food plant. It provided a reliable source of protein for the Aztec empire. Harvested from the many lakes and canals and made into a bread-like substance, "techiutlatl" was plentiful enough to be eaten every day. It was reported to be a food favored by soldiers and runners because of small bulk and concentrated nutrition.

Spirulina has also been gathered by the people who lived around Lake Chad for many centuries. The algal mats drift to the shores of the lake and its surrounding ponds. There the algae is collected in clay pots, rinsed through coarse cloth and set in the sun to dry. These blue green cakes are cut into squares for sale in the market. Rather than making the "dihe" into way bread as the Aztecs did the Kanembu people use spirulina as sauce to be poured over grain, vegetables or meat.

Birds, fish and insects feed on abundant algal blooms in many lakes. In the Rift Valley millions of flamingos feed on the spirulina in Lake Nakuru (Kenya) and Lake Kilotes (Ethiopia).

Spirulina is more tolerant of high salinity and high pH than most algae and can become the dominant species in lakes with rising levels of mineralization. Most modern spirulina projects are designed to take advantage of this quality.

Concentrated Nutrition Spirulina is 65 percent protein and amino acids, by far the highest of any natural food. The essential amino acids present are isoleucine, leucine, lysine, methionine, phenylalanine, threonine, tryptophan and valine. There are ten other valuable amino acids in spirulina including cystine. Spirulina's vitamin content is impressive. Spirulina contains Vitamin A, Vitamin D, Vitamin E, Vitamin K, riboflavin, thiamin, niacin, pyridoxine, folic acid, biotin, panthothenic acid and inositol. Spirulina is also high in minerals supplying calcium, iron, phosphorous, zinc, selenium, copper, manganese, chromium, sodium and potassium.

Spirulina is an effective source of iron and is often recommended to cure anemia.

Interest in spirulina as a modern food started in 1940 when the French phycologist Dangeard reported spirulina use around Lake Chad. Twenty five years later a Trans-Sahara expedition confirmed the presence of edible algal blooms there which were being harvested by local people. Botanists and other specialists in the scientific community became interested and many methods of cultivating and processing spirulina were tried all over the world. (Henrikson, 1997, pg. 17)

Modern Projects Some modern projects have been built on the shores of alkaline lakes to take advantage of the heavily mineralized water. The high pH keeps other organisms from invading the culture.

The spirulina project at Lake Chenghai in Yunnan, China is a good example of this kind of spirulina producing area.

Figure 12: Sirulina Raceway, Yunnan, China

Chenghai lake in isolated in Yunnan's high mountain ranges. The lake is too heavily mineralized to support fish though a species of brine shrimp thrives in its waters. This brackish water is pumped directly from the lake into a series of marvelously productive raceways built within glass houses. These houses allow for continued spirulina growth during the winter months. The annual productivity for the covered raceways is estimated at four kilos per square meter allowing for a forty-five to fifty day growth period before the spirulina is harvested.

Figure 13: Algae Drum Drying System

Spirulina rich water is collected in a central pool, sieved and drum dried on the premises. In Yunnan spirulina is made into cookies, bread, soft drinks, protein supplements, cosmetics, oils and ointments.

Photo by Elaine Solowey

Figure 14: Algae Drum Drying System

This huge modern project on the edge of a rather desolate mountain lake with little agriculture or animal husbandry possible in the area, points out how important spirulina production could be.

It is possible to produce immense amounts of protein food in a small area, 20 times more protein per acre than soybeans, 40 times more than maize.

The amount of water needed to produce spirulina is also much less. One kilo of spirulina requires the use of 2,100 liters of brackish water. Production of a kilo of soybeans requires 9000 liters of fresh water. Production of one kilo of US hybrid corn requires 12,500 liters of water.

Another interesting modern project is underway: **Earthrise Farms** in California's hot Imperial valley. Founded in 1982 this is currently the world's largest spirulina operation producing well over 500 tons of spirulina powder per year. Unlike the Chenghai project the raceways at **Earthrise Farms** are not covered. The pond cultures are kept pure by balancing the pond ecology at a high pH. From April to October the ponds are harvested every day around the clock to keep up with the explosive growth rates. During this period the spirulina blooms can double their biomass in two days.

The **Earthrise** operation is not as dependent on hand labor as the project at Lake Chenghai. Water from the raceways is pumped through a series of screens and filters. The nutrient rich water is returned to the ponds and the spirulina is separated out for drying without being touched by human hands.

The drying method at **Earthrise** is almost identical to the spray drying done at Lake Chenghai. Spirulina droplets are sprayed into a drying chamber to flash evaporate the water. The dry powder is exposed to heat as it falls to the bottom of the drum. Then it is collected and packaged

Current Use: Spirulina is still very much a "health food" or a supplement in the West. Over forty countries are experimenting with spirulina on large and small scales. Some large scale projects have gone over to the cultivation of "easier" algae, others have gone out of business because of technical problems but the large scale cultivation operations that remain are very profitable. Hundreds of small scale projects have been started in the developing world from "family sized" production units in India to the seeding of naturally alkaline ponds in Myanmar. India and China, the world's most populous nations, have recognized spirulina as "a national food resource."

Potential Spirulina can be grown in tubes, bioreactors and microfarms making it a promising element for home production, permaculture, food production in space and microgardening systems. Indeed, a truly efficient bioreactor for spirulina cultivation has become something of a Holy Grail for agricultural engineers.

Spirulina consumption by humans will probably be limited to a few super-nourishing tablespoons a day but spirulina can make up 33 to 35 percent of the diet of domestic fowl, farmed fish and shellfish. In other words we can feed spirulina to birds and fish and then eat them, even if we cannot consume large amounts of spirulina directly.

Comment The most exciting protein crop of the century which could be very important in the protein poor countries of the world. Spirulina may have extremely valuable medicinal properties. It may be the first crop cultivated out of the biosphere. Spirulina could nourish astronauts on the way to Mars.

Spirulina In Space

Spirulina is usually not taken seriously as a new crop, relegated to the status of "health food" or "food supplement." Few scientists believe that spirulina, as useful and as nutritious as it is, will be anything but a minor crop or a good source of powdered protein.

In one proposed agricultural system, however, spirulina is the food that will allow mankind to reach the stars. (Savage, 1994 pg. 45-48)

In 1994 Marshal T. Savage published a book called *The Millennial Project: Colonizing the Galaxy In Eight Easy Steps.*

Whether the work is regarded as futurology or as pure science fiction it is an ambitious and riveting work combining an audacious imagination with a dozen disciplines to outline the "steps" which may take mankind into space.

The first step is the realization that man's destiny is in space and the establishment of an organization to spread the word.

The second step would be to heal the earth by establishing colonies on the sea which would pump up great quantities of nutrient rich water by the use of the Ocean Thermal Energy Conversion (OTEC) system. The OTEC system produces power from the temperature difference between the cold water in the depths and the warmth in the top layers of seawater. The artificial upwelling zones would soon teem with life as the base of the marine food chain increases. The sea colonies would live off this bounty, all the while accreting the very stuff they are built out of from the minerals in the ocean through the judicious use of OTEC's nonpolluting energies. The colonies would relieve the population pressure of earth's overcrowded land masses and fight global warming by cooling down the water at the equator since water vapor is one of the contributing factors of the Greenhouse effect.

Once the sea colonies are established, a great effort would be made to establish colonies in space between the earth and the moon (step 3). Nested bubbles full of plants and colonists and shielded by water would bloom in the void (step 4). There people would adapt to space and new technologies would be developed to help colonize the moon and terraform Mars (steps 5 and 6).

After the moon is covered with domed ecospheres and Mars is terraformed the asteroid belt is next (step 7). Until we escape our solar system and take our place as a galactic life form (step 8). (Savage 1994, pg. 610)

These are exciting ideas and certainly more appealing than waiting until overpopulation, mismanagement of the earth's resources and human shortsightedness cause our civilization to collapse.

Empires rise and fall on their agricultural systems as is quite apparent from the fates of the Mayans and the Babylonians.

Spirulina cultivation is seen as the mainstay of a solar, then a galactic civilization. The algae will be cultivated in tubes, panels and other areas exposed to light, fertilized with wastes and recycled water and turned into a variety of processed spirulina based foodstuffs flavored with biotech flavanoids.

There are other good reasons why this scenario will probable remain science fiction, such as the fact that humans do not deal well with prolonged weightlessness, but certainly a spirulina based diet does not seem healthy or even possible. The proposed expeditions to Mars may include both spirulina and dunaliella cultures. But they will also be carrying generous amounts of conventional food to nourish the crew.

Mankind may indeed live in space someday, but if so there will have to be something more than single-celled wonders on the menu.

Dunaliella: Orange Spinach

Dunaliella is a marine algae that thrives under extremely saline conditions. It grows naturally in the Dead Sea and Utah's Great Salt Lake. Mass cultivation of algae requires large flat tracts of land where solar light is plentiful and temperatures are warm to moderate throughout the year. Such conditions exist in deserts and semi-arid lands. Fresh water is usually scarce in otherwise appropriate areas but saline water and sea-water are almost always available. Dunaliella grows best in media which contains between six percent and 12% salt which makes suitable for mass cultivation with water pumped directly from the sea.

Other Interesting Qualities Dunaliella is one of the few microorganisms which can survive in such saline surroundings. Other microorganisms are killed by the high salinity providing a selective advantage which allows for the cultivation of a relatively pure culture.

Dunaliella's optimal growth conditions discourage colonization by bacteria and fungus, the careful balancing of pond ecology necessary for spirulina production is not necessary with "orange spinach" though mixing and CO_2 introduction is necessary if the culture is to be productive.

Dunaliella lacks a rigid cell wall, making it digestible for humans. This algae has also been suggested as a possible "en route" crop for deep space voyages because of its rapid doubling of biomass and its high concentration of beta carotene. The proteins in dunaliella are similar in composition to other plant proteins such as soybean meal, being high in lysine but low in cysteine and methionine.

Glycerol can be extracted from dunaliella paste as well as beta carotene. Beta carotene, the more valuable product, can make up to 10 percent of the algal organic weight.

Dunaliella does not do well in bioreactors, tube gardens and sloping culture systems that are used to cultivate spirulina. The cells multiply too quickly and light cannot sufficiently penetrate the media.

Raceway cultivation has proved successful and dunalliella cultures are very productive, replacing some failed spirulina projects and producing a reliable supply of marketable extracts.

The dunaliella cultivation project in Israel's Gulf of Eilat is an example of a mass cultivation system that converted from spirulina production to the production of "orange spinach." The reasons for the switch are not hard to understand. Dunaliella could be cultivated with seawater pumped from the gulf of Eilat but did even better in "enhanced" seawater. These hyper saline waters are byproducts of reverse osmosis desalination and of salt production, both which are being done within a short distance of the dunaliella project. Dunaliella is less sensitive to the high temperatures of the summer months and the cultures are almost impervious to contamination since there are few organisms that can survive in such a saline environment.

Current Use: Cultivated for the production of beta carotene, glycerol, wastewater purification.

Potential Use: Ecological restoration projects, "out of biosphere" cultivation possibilities, nutriceuticals.

Comments: Dunaliella is difficult to filter because of its small size even though the algal mass is considerable. It does not form strands as spirulina does. New filtration techniques could make mass cultivation easier and more profitable.

Spirulina Cola Anyone?

The United Nations World Health Organization (WHO) in Geneva says about spirulina:

Spirulina represents an interesting food for multiple reasons and it is able to be administered to children without any risk. We at WHO *consider it a very suitable food.*

Spirulina might be considered a good food for children by the WHO but it is not much more attractive than powdered spinach—not exactly a favorite snack.

In China's Yunnan province however spirulina has become an important ingredient in the kind of food children love to eat. The blue green algae is baked into sweet green wafers called Lucky Wafers, added to steamed bread and made into a green bubbly drink called Lucky Cola.

At the Chenghai project in the mountains of Yunnan spirulina has been made into whiskey, green of course, with as much kick as whiskey made of more conventional substances. This last product is decidedly not for the kids.

Other spirulina treats from China include spirulina tea, spirulina chips flavored with green onions and spirulina noodles.

Haematococcus—Microagriculture in a Big Way

A pilot project has been established at Kibbutz Ketura in Israel's bleak Rift valley. (colorplate 26 on page 269) This arid and nearly waterless area is one of the hottest places on earth during the summer months. Establishing a huge bioreactor here, the first of its kind, for the cultivation of Haemotacoccus seems at first glance to be an act of folly. The large white concrete yard, surrounded by silver fencing with its rows and batteries of long glass tubes was quickly nicknamed "*the Kingdom of Glare*" by the locals. Surely a system with a proven tendency to overheat would boil over in the Arava's five months of relentless summer heat.

Photo by Ruthi Rubenstein

Figure 15: Green Stage of the Haematocuccus Project

Now in the pilot project's second season of operation high quality astraxanthin, a red pigment with considerable value in the production of fish food, cosmetics and nutraceuticals, is being produced on a large scale. The long banks of pipes are cooled by a mist of saline water and shaded by 60% shade net until the organism has multiplied sufficiently to be pumped out into the full sunlight. Stress turns the algae red and then the batteries of tubes are pumped out into the collection and drying apparatus.

The color plates (colorplate 23 on page 268 through colorplate 26 on page 269) should give some idea of the planning, investment and skill needed to succeed with this kind of cultivation.

What Else is Out There? I have described only a few organisms of a huge kingdom of single-celled wonders. Other algae are being cultivated commercially for other pigments, as elements of "living machines" and for animal food. There are many others which may produce food and useful compounds for us if we can learn to cultivate them and countless others which have not yet been investigated or described.

This is a task that must be undertaken if we are to explore all paths on the way to abundance.

Photo by Ruthi Rubenstein

Colorplate 18: Topaz Fruit

Photo by Elaine Solowey

Colorplate 19: Acacia Tree

Photo by Ruthi Rubenstein

Colorplate 20: Marula Grove

Photo by Elaine Solowey

Colorplate 21: Gathering Marula Fruit

Photo by Ruthi Rubenstein

Colorplate 22: Quandong

Photo by Ruthi Rubenstein

Colorplate 23: Algae before Drying

Photo by Ruthi Rubenstein

Colorplate 24: Blue Potato

Photo by Ruthi Rubenstein

Colorplate 25: Drying Algae into Flakes

Photo by Ruthi Rubenstein

Colorplate 26: Haematocuccus Project at Ketura

Photo by Elaine Solowey

Colorplate 27: Spirulina Raceway under Glass

Photo by Fritz Geral Schroder (Reprinted from The Growing Edge)

Colorplate 28: Plant Plane Hydroponics

XIV. The Plastic Cornucopia

Agriculture was invented to give the plants we depend on a competitive edge. We irrigate to allow the plants to flourish when there is insufficient moisture; we plow to make it easier for our crop plants to take root; we weed to keep out competitors.

The "perfect" agricultural system would be one in which every factor was controlled and only crop plants were introduced.

Modern society has attempted to introduce this kind of agriculture into the world's ecosystems causing an avalanche of negative consequences.

However there are many places where it has never been possible to grow food crops and "perfect" systems can be built from scratch.

New technologies and new materials have made agriculture possible in the Antarctic, in space, on the sides and roofs of high-rise buildings.

In these places "perfect systems" can be built and do little harm. In big cities it is even possible to get a good and profitable yield of perishables and herbs. Certainly this is wise way to use some of the dead spaces in the urban environment.

The following are constructed agricultural systems that to paraphrase a popular science fiction TV program "grow where no one has grown before."

Grow-Tech 2000: A Self-Contained Robotic Phytofarm

This interesting system was developed in Israel at the Technion entrepreneurial incubator by agricultural engineer Lior Hessel. It consists of stackable growth trays equipped with their own sensors, watering system and lights that fit inside a metal shipping container. It is an environmentally friendly system using small amounts of water, organic fertilizers and electric power. It does not produce large quantities of spent solutions and waste media as many soiless systems do.

Robotic sensors keep light, carbon dioxide, humidity and temperature at optimal levels. This allows for the rapid development of vegetables such as lettuce, cabbage, broccoli and herbs such as parsley which are easily damaged by weather and pests when grown in the open.

Hessel envisions Grow-Tech systems in supermarkets, restaurants and fast-food chains. He also hopes that his system will help compensate for Israel's lack of arable land and will conserve water that is now used in irrigating field crops. (Siegel, 1999)

Advantages: Fewer wastes than other soiless systems. Needs little space. Produces high quality perishables.

Disadvantages: Uses electric power from conventional sources. Very high initial expense. Requires skilled maintenance.

Evaluation: Good for big cities and developed nations that are short on arable land such as Israel and Singapore. Despite the predictions of the inventor it cannot revolutionize agriculture; it is simply too expensive.

CELSS: Growing Plants in Zero Gravity

Controlled Ecological Life Support Systems (CELSS) were designed by engineers Tuan Q. Cao and Thomas J. Slavin with technical support from Robert A. Anderson for the Boeing Company's Defense and Space Group.

CELSS are designed to deliver nutrient solutions to plants in zero gravity without letting the liquid drift away. This is achieved by constructing double plastic compartments connected by tubes carrying nutrient solution. The solution fills one compartment, which is separated from the other by a .45 micron Versapor membrane with a 41 micron mesh nylon screen. The plants' roots absorb nutrients through this layer and the liquid remains contained. The plants' roots are supported by a light perlite media and a negative pressure feedback system controls the circulatory pressure in the system. The oxygen level can be kept down in the chamber by a palladium catalytic oxidizer. This is necessary to allow for healthy plant growth and to reduce fire hazard aboard a spacecraft.

Nutrient Film Technique (NFT) has also been adapted for space at the National Aeronautic and Space Administration's (NASA) Ames Institute near Palo Alto, California.

Reprinted from The Growing Edge

Figure 16: NFT Hydroponics

Light will probably be supplied to CELSS by light-emitting diode (LED) technology. These appliances are sometimes called solid state lamps and supply a broad spectrum of useful available light.

While current systems are being designed to supply greens for the first interplanetary flight, other Ames researchers are working on a food system for a lunar or Martian base which would supply greens, fruits and carbohydrates, help purify the air and recycle wastes and water. Enclosed space is expensive to construct and pressurize. It is extremely costly to boost components for food systems out of the earth's atmosphere. So the ideal system will maximize space and minimize equipment. Ames researchers believe they can keep a person supplied with carbohydrates with ten square meters of growing space. They assume that another ten will be able to supply that lucky spaceperson with fruits, vegetables, herbs and even flowers. (Buchanan, 1995, pp 31-38)

Advantages: This system is modular in nature and can be set up in small awkward spaces. If people really are going to Mars and want to eat something besides algae this might be the only way to do it.

Disadvantages: Very expensive, computer controlled, only suitable for small plants so far.

Evaluation: For space stations and long space missions. No earthly applications so far.

Hydroponics: A Brief History

Hydroponics literally means "water-working." It is an agricultural method with ancient historical roots.

The first recorded use of soiless culture was in the famous "Hanging Gardens of Babylon" which were considered one of the wonders of the ancient world.

Muddy, rich waters were drawn from the river bottom and directed along stone conduits to the roots of the plants. The plants themselves were growing in shelves, niches and terraces, their roots covered with small stones.

There were hanging gardens in China and floating gardens in Mexico. Both systems were based on using silt to nourish the plants.

Theophrastus who lived from 372 to 287 BCE undertook various experiments in crop nutrition and isolated five substances he declared were necessary for plant life.

Dioscorides undertook botanical studies in the first century AD and made similar observations. He declared that plants could not live without water, air or light.

In 1600 a Belgian named Jan van Helmot showed that plants obtain much of their substance from water. He planted a five pound willow shoot in a barrel containing two hundred pounds of dried soil. After covering the barrel to keep out dust he watered it with rainwater for five years. The willow shoot gained 160 pounds while the soil lost less than two ounces. He did not realize, of course, that plants also derived substance from elements in the surrounding air.

In 1699 an Englishman named John Woodward grew plants in barrels containing various types and concentrations of soil. He concluded that plant growth was a result of substances in the water which are derived from soil.

In 1804 a French scientist named De Saussure proposed that plants are composed of chemicals obtained from water, soils and air.

In 1851 the French chemist Boussingault confirmed this and isolated some needed elements including nitrogen.

In 1860 and 1861 two German scientists by the names of Sachs and Knop eliminated soils from their experiment and grew plants in "nutri-culture" using mineral solutions.

This was the beginning of modern hydroponics. Since then soiless culture has split up into many schools and variations including the zero-gravity and Antarctic systems described in this chapter.

Why do hydroponics in the first place? The general advantages of hydroponic cultures.

1. Hydroponic systems can be tailored to the temperature requirements of the crop. A good example of this is cold bed agriculture in which the growing medium is cooled by subsurface pipes, the crop is irrigated by cold water or the entire green house is cooled by an airstream of the desired temperature, allowing valuable food plants to be grown in very hot areas.

Passively cooled systems are well suited to sub-irrigated gravel culture and have been used to grow crops as diverse as pitayas, apples, papayas and avocados in netted houses.

2. Hydroponics systems can be built around the spatial limitations of the site. Tomatoes can be grown in long vertical bags of media fitted with appropriate fixtures, greatly increasing the number of plants that can be grown in a cubic meter of space.

Cascade systems rely on descending zigzag tiers of tubing which can be stacked up to two and a half meters high, taking advantage of every inch of greenhouse space.

Igloo systems allow the plants to grow with their roots inside a Styrofoam dome. The curve of the dome increases surface area and more plants can be supported on it than on a straight square foam form.

In a hungry world of ten billion it might be necessary to produce food in small, cramped urban spaces. Hydroponic gardening can be done in *very* small spaces.

3. Hydroponics can be built around the limitations of the labor force. A flat, v-channel garden system can produce seven times the amount of cucumbers as an identically sized kitchen garden without hoeing and weeding or spraying.

Eggplants and strawberries in hanging gardens are ten times as productive as irrigated plants grown in the soil. Most of the labor and expense goes into establishing the system. But after a hydroponics system begins to work the amount of labor needed for maintenance is much less than that of an equivalent garden area.

4. Hydroponics systems can be tailored to the demand for specific types of produce. New crops can be started immediately after the harvest of the old. The systems can be sterilized with hot water, renewed and ready to be reseeded without lengthy sterilization periods or fallows. Successive cropping of seasonal produce by microclimate adjustment allows for year round production of fruit and vegetables.

5. Hydroponics systems are water efficient. Evaporation is minimized by covering channels, sumps and plenums and collecting greenhouse condensation. With areoponics and NFT very little water is allowed to evaporate. Most hydroponics systems are closed, recirculate and renew nutrient solutions through the life cycle of the crop plant and do not discard the solutions until harvest.

Gardening On Ice: Growth Chambers at the South Pole

Antarctica, which is covered with up to two miles of ice and about 50 degrees below zero, is not an easy place to set up a vegetable garden. But Antarctic stations support small long term human populations who are cut off from the world for many months, months in which they have no fresh food. The solution to this problem was the establishment of controlled environmental growth chambers which supply the galleys of ice stations with fresh herbs and vegetables.

The growth chamber at McMurdo Station is a good example of the controlled chamber system. It was established in 1988 when a volunteer work force refurbished dilapidated modular structures and joined them together.

A hydroponics system was set up with left over materials and supplies. Vermiculite was packed into short lengths of three inch polyvinyl chloride (PVC) tubes with a plastic screen across the bottom. The tube sections were seeded, then placed in trays that had nutrient solution running through them. When the seeds sprouted the plant's roots grew through the screen and into the nutrient solution while the tightly packed vermiculite supported the plant.

Artificial light is necessary since there is a dearth of usable solar radiation. The lighting in the McMurdo Station is provided by 400 watt High Pressure Sodium lamps on an

18 hour period. The inside of the growth chamber is lined with reflecting foil. Total power consumption is 14 kilowatts.

The walls of the growth chamber are insulated so it can actually get too hot inside. A ethylene glycol loop cycles coolant from an outside heat exchanger to an inner counterpart. Two fans help circulate and renew the air. The project also has a CO_2 enhancement system. Bottled $C0_2$ is released four times a day into the growth chamber.

Five years after the 480 meter square project was established it produced 1,025 pounds of lettuce, tomatoes, peppers and culinary herbs for the researchers and scientists who stayed through the “winter” months at McMurdo station.

When all the costs were figured in, produce grown at McMurdo station costs about four dollars a pound.

Similar projects have been established at other stations in Antarctica to supply fresh food and give the stations’ personnel a chance to see, smell and work with green plants. (Sadler, 1994, pp 36-39)

Advantages: Psychological boost for people in a barren environment. Fresh food that does not cost more than the produce which must be shipped in.

Disadvantages: Power is provided by diesel generator. There is no way to recycle spent solutions and media.

Evaluation: If people are going to live at the South and North Poles they should have fresh food and be able to see green plants. Worth the trouble just for life-quality enhancement. A better power supply should be found and way to cut down on waste materials.

Bioponics: The Application of Organic Gardening to Hydroponics

Bioponic systems try to combine the superior nutritive qualities of organically grown foods with the controlled environment of conventional hydroponic systems. This is done by growing the plants not in a media of water and chemical salts but in solutions inoculated with selected micro organisms to create a living environment for plant growth.

Bioponics uses a broader spectrum of trace elements and establishes a complex chemical reaction necessary to the manufacture and breakdown of plant acids. A mixed colony of microorganisms stimulates plant growth by creating a self regulating environment, avoiding the recurring imbalances and nutrient "spikes" of inorganic culture. The microbes also make up a "buffer zone" preventing pathogens from entering the host plant.

Microorganisms can be introduced to combat diseases or help correct deficiencies. Because of the living nature of bioponic solutions the dissolved oxygen levels of the media must be maintained at an 80 percent saturation factor. The bioponic solutions then mimic the biologically active environment found in soil. The high yields and pest free environments of classic hydroponics are retained. The resulting vegetables and fruits are of high quality in taste, appearance and nutritive value. (Thomas, 1991, pp 37-42)

Advantages: Bioponic produce can be considered organic. The taste and keeping qualities are good.

Disadvantages: It is not easy to inoculate solutions properly or maintain the desired oxygen saturation rate. Not for the amateur gardener.

Evaluation: For the organic gardener in environments when soil-based organic gardening is not possible.

Areoponics: The Mist System for Plant Growth

Reprinted from The Growing Edge

Figure 17: Home Aeroponics

Areoponics replace the growth solutions and media with a fine continuous spray of nutrient rich mist. The mist is usually contained in vertical growth fixtures which support the plants' greenery while the roots dangle down inside the tube. The mist in the tube must continually circulate lest the roots dry out.

The first areo-hydroponic system was developed in Israel by Dr. Hillel Soffer of the Volcani Institute to overcome problems presented by hot dry conditions at the test site at Ein Gedi. The system encouraged healthy root growth and was especially valuable in projects where plants were propagated from cuttings. The new method recirculated and reused the hydroponic solutions thereby reducing the pollution problem of commercial hydroponics.

Commercial kits have become popular for home areoponic gardens. A typical home kit is composed of a set of horizontal tubes with ring supports and plant sites, a mister, a nutrient solution sump and a circulation pump.

Florida's Epcot Center contains a large areoponic demonstration module where all kinds of vegetables and flowers are grown including large squashes which pass conveyor style through the nourishing mist. (Brooke, 1990, pp 25-26)

Advantages: There is no spent media such as rock wool or perlite to dispose of. Areoponics systems recirculate the solutions efficiently. It is spatially flexible.

Disadvantages: Root disease spreads with extraordinary speed through areoponic projects. Expensive to establish. Uses quite a bit of electricity.

Evaluation: To reach its full potential areoponics needs to be less vulnerable to disease.

Plant Plane Hydroponics: Hydroponics on Flat Surfaces

This intriguing system investigates the possibility of using a more environmentally friendly medium to do soiless culture (colorplate 28 on page 270). Developed in 1987 at the Institute of Vegetable Production in GroBerren in what was then East Germany this system uses a biologically based cloth instead of slabs and rows of rock wool or channels of nutrient solution. A thin layer of growing medium, sometimes jute fleece, peat paper or cloth made of rice husks, is spread over a plastic foundation sheet which seals off the greenhouse floor. White polyethylene is spread over the top to reduce evaporation of the nutrient solution and retard algal growth. A slope of one to two percent is necessary for a easy even flow of nutrient solution. Seedlings are started in small rock wool cubes then transplanted into the medium through slits cut in the polyethylene without cradles or ring fixtures. (Schroder, 1991, pp 62-55)

Advantages: Uses much less rockwool, does not need pumps as the system works on gravity, is versatile and the medium can be composted.

Disadvantages: Requires very careful floor preparation. Can only be done on the flat.

Evaluation: A good low cost system which does not use much energy.

Nutrient Film Technique: Water Thrifty Hydroponics

NFT systems circulate a shallow film of nutrient solution across covered growing surfaces. This allows for a substantial saving in medium, water and solutions but sometimes leads to root rot, arrested growth and other problems which are often due to the overheating of the root zone in a greenhouse environment.

A variation on the NFT systems deepens the film and insulates the growing surfaces in a hybrid between classic NFT and ordinary V-channel hydroponics. This allows for thermal stability which is sorely missing in ordinary NFT systems.

NFT's spatial versatility has made it the preferred system for zero gravity gardening. The enclosed growing surfaces can be deployed in many combinations of vertical and horizontal modules.

Advantages: Spatially adaptable, does not produce spent medium.

Disadvantages: Commercial systems use immense amounts of Styrofoam to anchor plants and cover the growth surfaces, not all of which can be reused.

Evaluation: Promising. Needs further refinement to be environmentally friendly.

Do hydroponic systems have a future?

In my opinion hydroponics may be an excellent way to bring food production back into cities. Even if the steady depopulation and degradation of the countryside is reversed there will not be a mass movement of urban dwellers to the rural areas. It has been estimated that in a world of ten billion people, seven billion of them will be urban dwellers. This means that a billion more people than

> the current population of the globe will be living in cities. (Brown et al, 1991, pp 97-109)
>
> Clearly it will be necessary to make cities friendlier places for people and the environment. The current urban landscape, where the dominant life-form appears to be the motor vehicle and people dwell uneasily in the spaces left over, is not a fit place to bring up the next generation of children. Urban gardening and urban hydroponics could help city dwellers reconnect to agriculture, provide jobs and work sites, encourage people to take up a healthy hobby and aid in the greening of the cities which must take place if they are ever to be made habitable.
>
> It is true that current hydroponics systems use too many artificial materials and too much energy to be considered sustainable. But the systems are being made more environmentally friendly and more economically sound all the time. Bioponics and hydro-organics systems are already producing vegetables and herbs. "Green" hydroponics will not be far behind.

We must not scorn the products of the "Plastic Cornucopia" if we are to achieve abundance.

Hydroponics may be the only gardening and food production option for extreme environments where conditions do not permit planting in soil.

Read more about different kinds of soiless gardening in

The Growing Edge magazine

Hydroponics for Everyone by Dr. S.K. Sutherland,

Hyland House, 1986

Hydroponic Home Food Gardens by Howard M. Resh, Woodbridge Press, 1995

Hydroponic Food Production by Howard M. Resh Ph.D., Woodbridge Press 1995

Any movement toward abundance in agriculture must be forward and not back. The agriculture that existed before "artificial manure," hybrid corn and crop protective chemicals cannot be restored. These systems would be inadequate now.

But there is no reason why the thrift, care and purity of the older systems cannot be combined with the tools and methods the 20th century has brought us for a new wise agriculture that will give us what we need yet be less of a burden on the earth.

There are many products, concepts and innovations that can be applied to sound and sustainable food production, from the thousands of kilometers of inexpensive plastic pipe used in drip irrigation to the compact computer technology of Global Satellite Imaging (GSI).

Simple farm tools such as shovels, hoes, high-pickers and wheel barrels should be made from the strongest, lightest and most durable materials. Farm machinery should be streamlined, scaled down, lightened up and keyed into renewable energy.

The lessons learned in the twentieth century workplace should be applied to agriculture. It has been found that most people do better work if they are part of a team and given some idea of the overall significance of what they are doing.

Industrial style farming has taken all the skill, knowledge, pride and joy out of raising food up out of the earth. The farm worker of today is not much better off than the 19th century worker in the sweatshop or the satanic mills: poorly paid, unappreciated, beset by occupational hazards.

I have never met anyone who wants to be a career farm worker. But I have met thousands of people who would like to be farmers. There is a vast difference between a tomato picker on a corporate agribusiness operation and a contributor to the immensely productive planned ecosystem that is a healthy organic farm. It is the difference between the assembly line worker and the skilled artisan.

When skilled labor returns to farming the agricultural systems will change again. A new sagacious agriculture will come to the fore, ecologically sensitive, more tuned to the seasons and the limits of the earth. These are some of the tools that can be applied to agriculture to help the wise farmer create abundance.

Drip Irrigation

This is the most important new system that has become available in the last 50 years because it saves, conserves and calibrates the most precious of all resources—fresh water.

Drip irrigation was invented in Israel's thirsty Negev by Simcha Blaas and put into commercial production by the Netafim company of Kibbutz Hatzerim. The original perforated hoses, simple line fixtures and plastic pipes have turned into an elaborate catalogue of drippers, misters, sprayers, pulse fixtures, water spikes and trickle valves which can irrigate almost any crop on any surface. Most drip "nets" are attached to a water clock or a small computer. Water is brought directly to the roots of the plant in exact quantities. There is no run-off and if the soil is mulched there is almost no loss to evaporation. There is also very little water-logging which slows the rise and accumulation of salts that has destroyed the fertility of so much irrigated land. (Clemings, 1996, pp 204-213)

Advantages: Easy to maintain and repair, versatile, thrifty of water and exact in water delivery.

Disadvantages: Serious initial investment, needs more care and repair with heavily mineralized water, wildlife may damage or gnaw pipes to get water.

Evaluation: The most important water saving system ever invented. Can reduce water expended on irrigation by 75 percent while also reducing evaporation and salinization. Can be combined with simple computer technology to make truly smart irrigation systems.

Integrated Pest Management

IPM is a concept rather than a product which combines the best of the old methods of pest control with the best of the new. Pests are "managed," basically outsmarted by a wide variety of strategies which keep the pest population low.

IPM methods include:

Companion Planting:

Planting vulnerable crops next to plants which have insecticidal or repellent properties. Much more is now known about which plants can be paired up for protection due to the work of the permaculturalists, both traditional and contemporary.

Target Materials:

The following are non-toxic substances which kill or repel insects in their most vulnerable stages of life: soap, vinegar, chili pepper, garlic, comfrey, calendula, diatomic earth and neem. Phyto-pesticides are also used to target specific pests. None of these substances is used for broad spectrum spraying so resistance does not develop in a pest population.

Biological Control:

Old methods of predator and pollinator introduction include catching ladybugs for release, planting "bee" plants in hedgerows and borders, building birdhouses and encouraging bats to take up residence near the farm. New methods of biological control offer an impressive list of pest-eating insects which are being propagated for sale to farmers and gardeners. For example, Nichols Garden Nursery in Oregon offers lady bugs, beneficial nematodes and Bacillus thuringiencis (Bt) for pest control, the Mason bee for pollination and mycorrhiza inoculants for soil and plant health. On the other side of the world BioBee of Kibbutz Sde-Eliahu of Israel offers a predatory mite for the control of spider mites, a parasitic wasp for the control of leaf miner flies, both a wasp and a gall midge for the control of aphids, predatory bugs for the control of thrips, a wasp to control whitefly and a wasp and beetle to control mealy bugs. Their favorite pollinator, sold in colorful cartons and delivered by messenger within Israel, is the desert bumblebee.

Many large nurseries and garden centers offer target materials, phyto-pesticides and insects for biological control.

Traps, Decoys and Pheromones:

A smorgasbord of sticky irresistible insect traps are available. They have usually been baited with food scent decoys or pheromones and are especially effective at trapping insects in the fertile stages of their lives. Some of these traps contain poison in the trap chamber but the dangerous material is not sprayed or dispersed in any way. Instead the insect crawls in to look for it. Since baits are species-specific this is an ecologically friendly way to whittle down the population of harmful insects.

Advantages: Flexible methods which do not allow large pest populations to establish themselves or develop resistance, non-toxic, low tech, low input. Pest populations decline and stabilize at much lower levels.

Disadvantages: These systems are not for absentee corporate owners, bottom of the barrel migrant workers and if-we-don't-spray-we-don't-eat chemical farmers. You actually have to know what you are doing to do successful IPM and you must have some connection with and knowledge of the land.

Evaluation: The best method to stop the pesticide spiral and the frightening rise in crop loss to pests which is happening despite all the poisons in the current system.

Nutrient Cycle Innovations

In the 1880s "artificial manure" became the great hope of agriculture simply because there appeared to be no other way to cycle enough minerals into the depleted soils of the Old World.

Unfortunately chemical fertilizers only appeared to solve this problem. The soil's demineralization has continued despite the chemical fixes used to drag crops out of the ground. Some foods contain a mere fraction of the minerals and vitamins they contained sixty years ago (Jensen, 1990, pp 4-20). Soil erosion continues at a frightening pace, especially in the "breadbaskets" of the developed world.

To remineralize the soil we must compost all agricultural waste products and get them back in the nutrient cycle. This includes human and animal waste which can be recycled in a way that does not spread disease. To put our own wastes back into the nutrient cycle in a safe, profitable and aesthetic way requires careful planning.

Sewer sludge from the developed nations is usually contaminated with pesticides, detergents and heavy metals. Garbage dumps are filled with plastic, paper, metal and glass which could be recycled along with millions of tons of food scraps which could be put back into the land.

Fortunately methods and products have become available to help us compost 90 percent of the wastes. These include cellulose digesters, bioinoculants, wood chippers, hot and deep composting methods and "living machines."

"Living machines" are engineered ecosystems for purifying extremely dirty water. They work well with sewer water from residential districts since these waste waters do not contain much truly toxic material. Designers of living machines have condensed the purifying powers of natural riparian zones into a few hundred meters of pools, ponds, aeration zones and settling basins. Water plants and bacteria are the living components of the system and the result is a flourishing green zone that produces usable water. Some of the larger systems have become recreational areas, something which would be impossible for a conventional sewage purification plant.

This is a description of the "living machine" of Arcata, California:

It looks like a vast park, but it purifies the town's daily batch of approximately two and a half million gallons of raw sewage. When you can turn sewage into scenery, you're on to something: it draws admirers from as far away as Japan.

On the surface the Arcata sewage treatment system is 150 acres of wetlands park at the edge of Humboldt Bay which empties into the Pacific Ocean. A sanctuary for people as well as wildlife, the park includes meandering streams, cattail covered estuaries, and freshwater ponds, edged with groomed paths and park benches. (Wallace, 1994, pg. 75)

Advantages: Truly an elegant cost effective way to deal with gray water, produces usable water, green spaces, is not offensive.

Disadvantages: Large projects need space. These systems are expensive to establish and must be planned carefully.

Evaluation: The only practical method so far of dealing with urban waste instead of dumping or ignoring it.

21st Century Compost: Straw Into Gold

A heroine in a ancient fairy tale can spin straw into gold. That ability pales beside the skills of a person who can turn weeds and garbage into money.

The new urban composter puts grass clippings, newspapers, food wastes, wood chips, leaves and weeds in a sleek bin, adds inoculants and water and is rewarded by rich compost that would cost twenty dollars a sack if bought from a commercial nursery.

This clever city gardener has also saved money on rubbish and garbage pick up and the materials needed to accomplish this can be found in most nurseries and gardening shops.

The large scale rural composter is paid by a municipality for accepting a city's food and garden waste. Rural composters are also paid to mow wild grass, cut brush and collect roadside vegetation. Since weeds are one of the best sources of trace elements they end up in the compost too. When the compost is bagged and sold the composter is paid again, this time for the product, something which rarely happens even in fairy tales.

Advantages: Raw materials do not have to be bought, they are all around us. The finished product is quite valuable. Solves waste problems and decreases fire hazards. For the most part composting does not need infrastructure or special equipment.

Disadvantages: Can be messy and generate pests if not done properly. Large projects require space and should not be too close to residential areas.

Evaluation: The most inexpensive way to remineralize soil, solution for the waste problems in our backyards.

Gifts from the Dust of the Earth

In the Biblical account of creation the first man was made by God from the dust of the earth. (Bible, Genesis 2:7) For some this is literal truth, others see it as a metaphor for our dependence on the earth for all our food.

In a sense we are continually being created and recreated out of the dust as we nourish ourselves from foods grown in the soil, have children and raise and feed our families.

The concept of grinding rock into fine dust and spreading it on ailing farm lands was developed by Dr. Julius Hensel in the 1980s. This stone meal, applied in amounts of three tons per acre, is capable of reviving sterile topsoil and regenerating dying forests. (Jensen, 1990, pp 74-76).

Rodale Institute, one of the first American institutions to teach modern organic agriculture, has published exciting material on the experimental use of rock dust. (Rodale Institute, 1995, pg. 25) This appears to be one of the more promising methods of remineralizing depleted lands since the technology is simple and the raw materials are common.

Advantages: A safe method of remineralization, the raw material is inexpensive, the results are immediate.

Disadvantages: Requires sturdy, expensive machinery. Rock dust is heavy and not easy to spread.

Evaluation: Especially suitable for forests, orchards and sloping areas where conventional fertilizer runs off and the long-term nutrient cycle must be considered.

Nutrients from the Sea

To borrow another verse from the Bible, "all the rivers run to the sea and the sea is not full" (Bible, Ecclesiastes 1:7) But not surprisingly that is where all the minerals end up. There are millions of tons of trace elements in seawater that are rare on land. This is why sea-based nutrient products are so effective in depleted topsoil. Foliar sprays based on seaweed are also potent and beneficial. Sea-based nutrients have many advantages over petrochemical fertilizers; they are cheaper, they are non-toxic and non-polluting, easy to apply and the supply is inexhaustible.

Advantages: Most preparations are liquid and quite concentrated. Sea based micro-element fertilizers are generally much cheaper than petrochemical trace element compounds. They are also non-toxic.

Disadvantages: Foliar sprays wash off easily in the rain.

Evaluation: An excellent way to deliver trace elements to farms and gardens.

Diversity Planting

Natural stands of plants are made of individuals each one slightly different. Monocrops are basically the same plant copied over and over and planted in huge areas. Diversity planting puts modern cultivars in an order closer to that of a natural plant stand.

Associated Press writer Jeff Barnard wrote an article on a Chinese experiment in diversity planting.

Chinese farmers who abandoned the modern practice of planting a single type of rice in their paddies and adopted the more natural course of mixing varieties were rewarded with bigger harvests and no longer had to spray expensive fungicides.

"While the benefits of genetic diversity were known to Darwin, the study serves as an important reminder at a time when agriculture is increasingly looking to high-tech solutions," said Martin S. Wolfe of Wakelyns Agroforestry in Pressingfield England. *"This deceptively simple experiment deserves widespread attention, partly because of the principle it illustrates, partly because it may never be repeated on such a scale."* Wolfe wrote in a commentary on the study published in Thursday's issue of the journal Nature.

The article describes the change made in sticky rice cultivation as rows of the vulnerable cultivars were alternated with other rice varieties. Sticky rice brings higher prices but is susceptible to rice blast, a devastating fungus controlled by spraying.

In the experimental areas rice blast incidence fell by 94 per cent and the yield increased by 89 percent. Alternating rows of different rice varieties thwarted the spread of the rice blast.

Oregon State University plant pathologist C. Mundt commented: *One way of thinking about this monoculture would be kind of like a field of dry grass: Drop a match in it. There is nothing to stop the fire from moving through it. A mixed population is like a field of dry grass and wet grass. Drop a match there and it is going to be slowed down. It will burn a dry grass patch, then hit a wet one.*”

Though fifty per cent of the planting was vulnerable to the fungus the disease incidence dropped to less than 10 per cent of what it had been in a sticky rice monoculture. (Barnard, 2000)

Experimental diversity planting when applied to other crops such as wheat, apples, pitayas and peaches has cut the incidence of disease by over 80 percent and bolstered yield by 50 to 70 percent. It must be remembered that diversity planting requires no additional equipment, investment or infrastructure. It is a wiser way of planting, one which more closely imitates a wild pathosystem and which has significant benefits for the farmer and the plant.

Advantages: Healthier farms, orchards and gardens for no additional costs.

Disadvantages: Planting and harvesting must be done row by row.

Evaluation: Diversity planting should be tried in all cultivated crops, especially edible ones.

Pattern Rotation: Allelopathy

A more sophisticated and diverse form of crop rotation which reduces pest proliferation and the build up of soil diseases by rotating delicate and vulnerable crops with insect repelling crops, legumes and grains. It has been known to farmers for many years that some crops reduce populations of harmful insects. The decomposition of certain crop residues may release sulfides and phenolic acids that have antibiotic or antiseptic qualities. The study of the interaction between the chemical compounds produced by the roots of plants or by plant mulch is called allelochemics. A fallow season in the rotation allows the soil to restructure itself and microorganisms to complete soil purification. (Yeomans, 1991, pp 33-36)

Advantages: Outsmarts the bugs quite effectively. It is effective against a variety of soil borne diseases. Needs no special equipment or infrastructure.

Disadvantages: Not for the weekend farmer. Pattern rotation gives an unskilled farmer a chance to fail with five different crops instead of one.

Evaluation: The pest population in successful rotations is stable and minimal. There is little crop loss or damage. A very effective way to manage pests and increase the soil's fertility.

Resistance Breeding

Learning from wild pathosystems is what Raoul Robinson advocates in his book *Return to Resistance: Breeding Crops to Reduce Pesticide Dependence*. This modest but amazing work follows the epidemiology of potatoes, sugar cane, tropical corn and other important plants and explains how they can be made resistant to plant disease.

In Robinson's opinion pedigreed plant breeding is one of the fundamental errors of modern agriculture resulting in debilitated, static and vulnerable food crops that cannot protect themselves or adapt to new conditions. He explains exactly why this is so leading the reader through a series of diagrams that explains the concept behind the breeding process and delineates the strengths and weakness of two great schools of plant genetics: Mendelian and Biometric.

Robinson believes that plants have the ability to defend themselves against insects, a notion which has been supported by the work of entomologists and plant researchers.

According to Nora Underwood, a plant researcher at the University of California at Davis there is a chemical language of plants. For example wild tobacco plants interpret the chemical signals from a nearby insect damaged sagebrush and mount a chemical defense. *"It's in the best interest of the tobacco to eavesdrop." Scientists are learning to interpret this silent cacophony of chemical chatter. If researcher can translate the often airborne or volatile signals they may be able to enhance a plant's ability to protect itself and develop ways to reduce pesticides.* (Gewin, The Oregonian, August 15, 2001)

The wounding of a plant may trigger the formation of two hormone like substances that turn on a variety of plant defenses. A jasmonic-acid mediated response may be triggered by an attack of mites or caterpillars, a salicyclic acid mediated response might be brought about by an attack of fungi or bacteria.

Or the plant may release a bouquet of chemical substance that allow a predator insect to home in on the attackers.

Wild plants have active defense systems and the advantages of biodiversity to protect them while most cultivated plants are genetically similar and have lost their ability for the most part to defend themselves.

While researchers at U.C. Davis and Penn State believe that they can turn on the defense system of most crop plants, Raoul Robinson believes that defensive abilities were present in all our food plants and have only been recently bred out of them. For example wild cotton produced eight to 10 times the quantity of volatiles as the domestic varieties when they were each fed upon by caterpillars.

"Each time you select for size or quality, you might be losing genes for another function of the plant and a lot of times it's defense,"

said a Washington State university researcher named Bud Ryan.

Another researcher, a Mr. N. Lewis of the Agriculture Department, commented *"Let's say for example that plant production specialists figure out a fertility regime that gives out another bale of cotton.... You may have weakened the inherent plant protection mechanism and it may cost you a bale to protect it"* (Gewin, The Oregonian, August 15, 2001)

Robinson said it is possible to breed these natural protective mechanisms back into food plants.

The fascinating third section of the book aptly titled *Solutions* contains plans, techniques and suggestions for breeding resistant plants, cutting pesticide use to an absolute minimum and returning natural defenses to cultivated plants.

Advantages: Effective way to reintroduce resistance. Especially suitable for the breeding of ecotypical cultivars. The field resistance established is not easily matched by pests or parasites so the resistant plants retain their resistance for many years.

Disadvantages: Requires time, skill, challenges the fundamental assumptions of modern plant breeders.

Evaluation: This is also a humane, wise, knowledge-based solution for the pesticide spiral as different from the expensive, clumsy, tech-infatuated solutions of genetic engineering as it is possible to be.

It is this kind of solution we must seek if we are to achieve food security in the coming century.

The word "abundance" has three definitions in the Webster's Seventh New Collegiate Dictionary (Webster, 1965, pg. 4).

1. an ample or overflowing quality
2. wealth
3. relative degree of plentifulness

The first definition of the word hints at carelessness, an overflow of something usually means there is too much of it. There is an overflowing quality about conventional agriculture that is not healthy. Huge quantities of food are produced but the food is poor quality and often contaminated by toxic chemical residues. There is mass production of foodstuffs but along with this mass production there is mass destruction, mass exploitation, mass neglect and mass waste.

The second definition has the chill ring of cold hard cash. There is wealth and evidence of wealth all over the world. The five hundred richest persons on earth possess more goods, assets, land and capital than the poorest three billion souls combined. (Durning, 1992, pg. 21-29) Clearly what is needed is less wealth and more abundance.

The third definition is the true definition if we seek abundance in the world. A relative degree of plentifulness would mean that no child would go to sleep hungry; desperate farmers would not slash and burn their way across the remnants of Asian forests every autumn; displaced peasants would not crowd the shantytowns of the developing world looking for jobs that do not exist.

We would also not have to watch the steady and horrible destruction of the natural world as our air becomes unbreathable, our water undrinkable, our health and safety expendable.

What's Wrong with These Pictures?

The National Geographic of February 1970 contains a startling article. It is called *The Revolution In American Agriculture* and subtitled *More Food for our Multiplying **Millions.***

It was an anachronistic piece in 1970. Farmers drawl in dialect and wear bibbed overalls. The description of black sharecroppers who lost their jobs in the cotton fields sounds like something from *Gone With the Wind.*

Flattering descriptions of work being done in plant breeding and IPM share pages with equally flattering descriptions of some of the most troubling aspects of modern farming: chemical use, monocropping, industrial style animal husbandry, racing on technological treadmill and the elimination of all but the largest and the most successful farms.

The most dated element of the article is the depiction of the farm of the future. Drawn by artist David Meltzer with the guidance of United States Department Agriculture (USDA) specialists in shades of green, blue and gold this farm features high rise pastures for cattle, crops being grown under climate controlled domes, huge combines tilling, harvesting, spraying, an automated slaughterhouse where cattle go in on one side and processed beef is packed into bus-sized cylinders on the other. A few people, almost lost in the weird Astroturf ambiance are strolling or pushing buttons in the bubble on top of the farmhouse which looks like a Unidentified Flying Object with a back deck.

No one is actually doing any work. There is no sense of place. The landscape is so totally artificial that the pictured farm could be in any flat spot on earth.

The text underneath the picture on page 184 sent shivers down my spine. *Fields will be larger, with fewer trees, hedges, roadways. Machines will be bigger, more powerful and will be able to do more operations in fewer trips across the land. They'll be automated, even radio-controlled with closed circuit TV to let the operator sitting on a front porch monitor what is going on. It isn't difficult to visualize agricultural plots several miles long and a hundred feet wide. Equipment straddling the strip will roll on tracks or paved runways, swinging around at the end to work the adjacent plot without a wheel touch compacting the soil in the cultivated areas. Weather control may tame hailstorms and tornado dangers. Atomic power may supply power to level hills and provide irrigation water from the sea. Satellites and airplanes overhead will transmit readings enabling a farmer to spot diseases breaking out more surely than he could by walking through his fields.*

In essence, the future will be more of the same: fewer people growing more monocrops using more chemicals and energy in an increasingly disturbed and artificial landscape.

I shivered because that vision of the future seems more likely to lead to the collapse of our agricultural systems and mass starvation.

This vision of the future is thirty years old. Have we learned anything in the meantime?

Many farmers and researchers found that vision both dismaying and out of touch with agricultural reality.

A 1996 version of the "future farm" circulated on the internet was even stranger and more out of touch.

Immense tracts of land would be sprayed, fertilized and observed by guided missile, plowed, harvested and planted by satellite controlled machinery. Huge combines would rumble unhampered around fields the size of entire states doing everything. *"And not a bibbed overall in sight."*

This of course is the key statement. No farmers need apply. The technological horn of plenty will keep the world fed without them.

The amount of money such a system would cost to establish and maintain is not mentioned nor is the possibility that computer error might cause Kansas to planted with pineapples and Denver to be plowed for wheat.

What the agricultural systems of the future will be like I cannot say. But the elements and practices that must be included if we are to achieve abundance are slowly becoming apparent all over the globe.

It is already clear that the "farm of the future" can be nothing like the farm in either version.

What will the Agriculture of the Future be Like?

Farmers Doing the Farming The dream of agricultural production without producers is an ignoble and impossible one. True abundance can only be achieved by skilled, proud and valued farmers and agriculturalists. It will be necessary to reverse the trend of the twentieth century and bring farmers back to the land. This is not as difficult as it sounds. Most displaced farmers left their land reluctantly and under great duress. When it becomes possible to make a living from the land again then farmers will return.

This must happen soon, though, lest the cultures surrounding agricultural activity vanish before the change can be made. This is the primary fear of farmer/philosopher Wendell Berry in his book *The Unsettling of America* and would mean that there would be no one left who actually knows how to live on and from the land. (Berry, 1977, pp 39-48) This would be a terrible loss. The farmers of the 21st century would truly have to start over again, not in a virgin wilderness but in a chemically desolated landscape.

Many countries have already tried to produce food without farmers. The Soviet Union eliminated and displaced its most skillful and successful peasants and replaced them with the unskilled, the unpaid and the unmotivated. The result is that now after the breakdown of the Soviet empire an area with one third of all the arable land in the world has become the world's largest grain importer. There is less land under cultivation in the former Soviet Union now than there was under the Czar. The people who should be farming the land to feed their fellow citizens are raising fish in their bathtubs, rabbits on their balconies and vegetables in tiny plots that they can at least call their own, not for sale but for survival.

The west was not much kinder to its farmers. They were driven off the land by various economic measures and crises but most of all because government help, incentives, subsidies and tax breaks went to the "efficient" farmer. Unfortunately "efficiency" usually meant monocropping, chemical use, machine dependency and the tendency to exploit the cheapest sources of labor available. To this day the governments of Canada, Brazil and the United States promote continuous monocroppings and discourage crop rotations, intercropping, organic agriculture and the use of alternative energy in farming.

Listen to the pain and sorrow in the words of Verlyn Klinkenborg whose editorial appeared in the September. 3, 2000 edition of the New York Times, describing how Iowa's population has aged and diminished and the number of farms has been halved.

> *It can be hard to grasp the implications of a shift this large. Think of a farm for a moment, not as a tract of land or so many acres of corn*

and soybeans. Think of a farm instead as a constellation of people, each of whom has ties to the community—to churches and schools, to banks and farm implement dealers, to co-ops and Lion's clubs. In 1960 small towns were thriving not on the prosperity of town residents alone but on the prosperity—and the population—of farms in the adjoining countryside.

Now half those farms are gone, their groves cut down, their houses bull-dozed, their constellations of people dispersed. Year by year the size of farms increases and their work force diminishes. The towns are very quiet, except the ones lucky enough to hear the roar of enormous grain driers at the elevator outside the railroad tracks. School districts have consolidated and small town business is just a ghost of itself. It is not uncommon to come across a sense of disinheritance among Iowans, especially those living in small towns. The bond that tied them to the farmland around them—a bond that was social as well as economic—has to a striking degree been severed. There is no one on the farms and the towns have no one to serve except their aging residents.

All this has come about because of a faith—an unproved faith—that the only way for agriculture to prosper is for farms to become bigger and bigger. There is no question that Iowa raises vastly more corn and soybeans than it did in 1960. But at what cost? Farm prices are as low as they have ever been and farmers now pray for emergency aid from Congress they way they pray for rain. One of the indisputable results of large-scale corporate farming has been the depopulation of the countryside. (Klinkenborg, 2000)

To bring the farmers back to the land governments will have to learn to value local expertise and local labor.

Agriculture must be made safer (currently farming is the third most dangerous profession), more profitable and more respectable for the people actually doing the work.

Discarding the Machine Ideal A pig is not a machine for producing pork and piglets. A chicken is not a machine for producing eggs. To treat sentient creatures as if they are machines is a philosophy of unparalleled stupidity. When the animals are treated in such a manner they live short, wretched, unhealthy lives and our health is affected by their suffering. The antibiotics needed to keep factory-farm animals alive are the most likely cause of the rise of antibiotic resistant bacteria. Beef and milk saturated with hormones has lowered the age of puberty and may be the cause of several types of human cancer. (Wolfson, 1998 pg. 3) Milk produced with recombinant bovine growth hormone (rBGH) has been declared unsafe by the United Nations for much the same reasons. (NPI Center, 1999)

Antibiotics, hormones and rBGH would not be necessary if animals were allowed some opportunity to exercise their natural instincts. It is only logical that an animal that is allowed to move about, groom itself, interact with other animals, breathe clean air and see the sunlight will be healthier than one that lives its entire life in a tiny pen or a crowded feedlot.

Needless to say farm workers who are treated like machines lose their health, their wholeness and/or their sanity.

What We Eat There is absolutely no reason why the rest of the world should aspire to the modern western diet of fats, sugars, processed carbohydrates and dubious proteins. Even if it were possible to produce the amount of beef, corn, white flour, sugar and chemical flavoring necessary to feed the world in this manner it *should not be done.*

This diet is a recipe for an unhealthy life, a debilitated old age and often enough an early death.

The food we eat should build our health and strength, give us the energy to grow, work, learn and raise our families.

We should not have to be afraid when we taste a grape or eat a slice of tofu that what we are putting in our mouths has been coated with poison, exposed to radiation or tampered with on a molecular level.

We should also not have to fear that the governments that we elected to represent us are systematically attempting to undermine the only power we have over what we eat: the power of free, informed choice. All irradiated food and GM food must be clearly labeled so that we can decide—not whether it is safe or not—but if we want to buy it and eat it.

Any attempts to pretend that irradiated food is equal to cooked food or that GM food is not different from its unmodified counterpart should be regarded as fraud.

Land and Community When the family farms vanished in the US, the small towns that had been supported by farming communities went into decline. This led to the disappearance of most of the last vestiges of true communities in the United States. It is possible for communities to form based on other shared interests. But the sense of belonging to one place and of living, working, and learning with people one knows is a basic human need. When this need is denied there is much suffering.

In the rest of the world the destruction of traditional communities has not been total. There is something to build on, a sense of nostalgia, a place called home, sometimes an actual village to go back to. When the essential connections between the community and the land are validated, small communities regenerate and reform.

Many people are acutely distressed by the noise, crowds, crime, and dirt of urban centers and live in cities not because they want to but because they must in order to support themselves.

Community supported agriculture (CSA) could be a powerful tool in the redemption of the village and the town. CSA will allow communities to reclaim the farmland of the world for sane, clean and sustainable agriculture.

Children At Risk A remarkable article published in the June 19th, 2000 edition of *US News & World Report* links many birth defects and developmental problems to our chemically saturated lifestyle.

A growing body of evidence suggests that neurotoxicants, a class of compounds found in vaccines, head-lice shampoos, paint, cleaning products, pesticides, herbicides and tuna are the significant contributing factor to the crippling of both body and mind of many children.

One in every six American children suffers from problems such as autism, aggression, dyslexia and attention deficit hyperactivity disorder. In California reported cases of autism rose 210 percent from 1987 to 1998.

The House Committee on Science and Technology reported that there were 850 known neurotoxicants *any which may result in devastating neurological or psychiatric disorders that impair the quality of life, cripple and potentially reduce the highest intellect to a vegetative state.*

The polychlorinated biphenyls (PCBs) unleashed by Monsanto in the sixties are among the 850 chemicals, as are the popular organophosphates Dursban and Diazinon.

The PCBs banned by the Environmental Protection Agency (EPA) in 1977, continue to haunt children, not only increasing the risk of cancer but impairing the memory and intellect of those exposed.

It is now known that Dursban can damage the brain. Diazinon, it is suspected, can ravage the vision and the immune system.

There are over eight hundred more neurotoxicants lurking in the chemically treated timber of the back deck, in the water that the child wades in, on the tomatoes and grapes and apples that the child eats until the concerned parent wonders if there is anything left in the world that is safe and wholesome to eat or any clean place left to live.

Lawsuits, superfund activities and city-wide cleanups are all necessary to end the chemical madness but for the children whose lives have been blighted it is too late. They will never be what God meant them to be, not as bright or as well-formed or as happy.

Their wholeness was sacrificed for what?

Certainly not to feed the world. Because the damage has been done and world is still hungry.

What kind of world do we want? I was fond of imaginative play as a child and sometimes I would say to my playmates "This is my story, I get to say how it comes out."

I have quoted the visionaries of Monsanto, the prophets of agribusiness and the messiahs of GM foods.

But this is my book. So now it is my turn to put forth my vision of the future.

Small vigorous communities are found everywhere, fitting into the most challenging environments, providing a thousand new frontiers for those who want to be independent. There is work for everyone who is willing. The five hundred new cultivars of the Second Domestication have ensured that safe wholesome food is available for all. Light, strong, solar-powered agricultural aids skitter over the rich earth, drilling seeds, cultivating, carrying fruit and grain as farmers gather the abundance. Wise agriculture is taught in schools and farmers are once more among the most respected people of the community.

It is possible to stop in any field and taste a grape or a tomato or a cucumber without ingesting anything worse than a stray ant or aphid.

Cows are in the pasture and chickens peck in the yard. They still end up on the menu but they aren't in misery for every moment of their existence until they do.

Global warming has been stopped by the banning of fossil fuels and a global tree-planting campaign.

No more plant or animal species have become extinct. The sea has cleansed itself.

The air is like crystal and the rain is like wine.

No child is crushed on a roaring road because a ball bounced into the traffic, the vehicles on the narrow winding roads, also light, sun-powered and used mostly for visiting, travel at leisurely speeds. No one is in a hurry because work, recreation, medical care, a place to buy and sell produce and hand crafted goods are all within walking distance of where a person lives.

Though ATV's and power skis have been banned there is still danger and excitement for those who wish to seek it. White-water kayaking, mountain climbing, and backpacking are popular among the young peo-

ple. There are ultralight planes and gliders in the sky powered by the sun and wind, as fast and as sleek as birds. Hunting with a .22 or a bow and arrow is also popular in designated areas. The hunters chase and track the animal on foot and face it alone. No real hunters miss the days when deer where slaughtered by automatic weapons after being chased down with a helicopter.

Luxury goods are expensive but durable, lasting for the purchaser's lifetime.

There are still computers and televisions but the user chooses exactly what he or she wants to let into the home.

Music, crafts, and the homey arts such as cooking, baking and brewing flourish. No one skips a good meal and settles for junk food. Most communities are graced with a choir or a drama group but all have access to the theater, music and literature of the larger society.

We still get old, we still lose loved ones, we still become ill and we still die.

But we know there will be food and water, work, love and solace for our children and grandchildren.

And we know that we have been wise, thrifty, productive and kind during our short time on earth.

What we decided we didn't need to know In the twentieth century much of humanity unlearned all the virtues that insured our survival as farmers. We discarded thrift. We have no patience with patience. We denied that we are part of the natural world.

We tried to create our own ecosystems in which nothing exists but the plants and the corporation that owns them. There was no longer any room for foxes and hedgerows and bugs and birds and butterflies. Very soon there was no longer room for the farmer and the hired hands, either and they were replaced by combines that are as elaborate and costly as jet planes.

All this effort was expended to make sure that the systems we created did not "leak" that no pesky creatures that did not plow and sow and so were not entitled to reap did not tap into our bounty. The knowledge that we were tapping into a bounty that we did not create, but only modified, totally left our consciousness.

A hundred years ago a sagacious soul might have said: "When you throw nature out the door she comes in through the window."

The modern adage will serve just as well: *What goes around, comes around.*

We discarded thrift and are faced with rivers of sewage, mountains of garbage, poisoned water, soil that is blowing away and a hazardous waste disposal problem of frightening proportions.

We dare not fight "progress" even if the process is taking us someplace we do not want to go. We cannot wait for the slow buildup of health and fertility that allows a piece of land to produce good food. So conserving measures are not "profitable" even though they are the only way to produce food without undermining the land's ability to support us.

We pretend we are not a part of the natural world and see the earth begin to disintegrate around us.

We make farms that have as much to do with reality as children's play gardens of matchsticks and plucked flowers lined up in the sand.

The insects we turned into superbugs lunch happily on our defenseless identical poisoned plants. We eat the same crops and sicken because the insects are stronger than we are and endlessly adaptable.

That is what we forgot during the twentieth century.

What must we learn to survive the 21st?

Avoiding Shortcuts There is no shortcut to feeding the world. The belief that scientific agriculture is going to save the starving is widespread. Unfortunately it is totally false because the mindset and practices that come with industrial style agriculture are not part of the solution; they are a *big* part of the problem.

We must understand that we are dependent for every molecule of food that we eat on other living creatures. The best way to safeguard ourselves is to care for the creatures that we need and the creatures around us. If we are dependent on plants and animals then for our own sakes we must not let them live in misery or degenerate or become extinct. Nor must we lose sight of the fact that they are living organisms with a value beyond that of what they fetch per pound.

The plants and animals themselves are dependent on the fertility of the soil, the benevolence of the climate, the blessing of sun and rain.

So we must insure that the soil is fertile and alive. We must insure that the sunlight continues to be filtered by the ozone layer, that the globe does not cook in a layer of C02 and the rain can fall without bringing down acid with it.

We must do all these things not for romantic or utopian notions about stewardship and ecology, but because we must if we want to survive.

The danger to us increases as the world becomes more crowded, more dirty, with each tree felled and each child born.

We cannot go back to the agriculture that existed before the industrial revolution. There are too many of us and too much has changed. If we tried to turn the clock backwards we would be as out of place as Rip Van Winkle.

We must go through to something new, to a kind of agriculture in which we use all the human qualities that were jettisoned in pursuit of our current false plenty: intelligence, patience, skill, foresight, thriftiness, cooperation and a deep and abiding respect for the beauty, diversity and fragility of the earth.

Our only home.

That is the way we will achieve true and lasting abundance.

Associated Press (AP) (2000) “Blair Concerned for Modified Food”, www.connectotel.com/gmfood/ February 27

Barnard, J (2000) “Study: Not fooling with Mother Nature brings better rice crop”, *Associated Press* printed in *Corvallis (Oregon) Gazette-Times pg.* 6.

BBC News (1999) “Farmers Welcome halt of terminator”, www.bbc.com xv, October 5

Bennett, J and George, S (1987) *The Hunger Machine,* Polity Press, UK pp 2,6,9,15

Bible, The, Hosea 8: 7 (whirlwind); Genesis 2: 7 (dust of the earth) Ecclesiastes 1: 7 (rivers)

Billard, J (1970) “The Revolution in American Agriculture: More food for our multiplying millions,” *The National Geographic Magazine*, vol. 137, no 2, pp 147-185

BOSTID (1996) *Lost Crops of Africa, Volume 2*, National Academy of Sciences Press, Washington DC, pp 39-112

Brasher, pg. (1999) “Anti-Biotech Activists Plan Lawsuits”, *Associated Press*, .www.netlink.de/gen/ZEITUNG/1999/990914B.htm

Brooke, L (1991) “From Desert, to Laboratory, to Backyard Aero-Hydroponics—The Hydroponic Method of the Future!”, *The Growing Edge*, vol. 2, no 1, Fall, pp 25-26

Brown, L et al (1991) *Saving the Planet*, W. W. Norton & Company, Worldwatch Institute, New York, pp 97-109

Buchanan, C (1995) “Zero Gravity Gardening,” The Growing Edge, vol. 6, no 3, Spring, pp 31-38

Cable News Network (CNN) (2000) “FDA investigates gene experiment for possible AIDS contamination,” www.connectotel.com/gmfood/, February 11

Carson, R (1962) *Silent Spring, Fawcett Publications, Inc., Greenwich, Connecticut, pp 217-231*

Clemings, R (1996) *Mirage: The False Promise of Desert Agriculture*, Sierra Club Books, San Francisco, pp 204-213

Coale, K (2000) “Anti-GE lawsuit against FDA has Clinton Administration worried,” www.purefood.org/GE/antigesuits.cfm, January 12

Cole, J (1979) *Amaranth from the past for the future*, The Rodale Press, Emmaus Pennsylvanis

Cohen, J (1963) Bernal Diaz: The Conquest of New Spain, Penguin Books, Great Britain, pp 215-232

Cohen, R (1999) “Fearful Over the Future, Europe Seizes on Food,” *The New York Times Weekly Review*, vol. CXLVI-No 51 as reprinted in *The Jerusalem Post*, August 29

Cummins, R and Lilliston, B (2000) "Genetic Engineering, Factory Farming, and Organics," Little Marais, Minnesota, www.inmotionmagazine.com/geff.html

Dawkins, K (1997*) Gene Wars, The Politics of Biotechnology,* Seven Stories Press, New York pgs 14-16

Dougherty, J (1999) "Seed technology: Feast or Famine? "Terminator" could be used to control food
supply, warn critics," www.WorldNetDaily.com, January 28

Durning, A (1992) How Much is Enough? W. W. Norton & Company, New York, pp 21-29

Foundation for Revitalisation of Local Health Traditions (FRLHT) (1995) "Neem Free No More," *A Green Health Campaign Series Bulletin 3*, FRLHT, Bangalore, India, pg. 13

Duveyrier (1864) quoted in BOSTID, (1996) *Lost Crops, Volume 2*, National Academy of Science Press, Washington DC, pg. 260

Fowler, C and Mooney pg. (1990) *Shattering: Food Politics and the Loss of Genetic Diversity,* The University of Arizona Press, Tucson, pg. 3

Frith, M and Murphy, E (1999) "Environmental Risks: How genetic engineering could break down our ecosystem," www.sage-intl.rg/genetic.html, May 17, pg. 6

Gardner, G and Halweil, B (2000) "The Underfed and Overfed," *World Watch*, vol. 13, no 4, July\August, pg. 25

Gillis, J (2000) "A Gift From Monsanto: Genetically Altered Rice," Washington Post Service reprinted in International Herald Tribune, August 5-6, pg. 4

Gewin, "A Shield of Green," The Oregonian, August 15, 2001

Goering, pg. et al (1993) *From the Ground up; Rethinking Industrial Agriculture*, International Society for Ecology and Culture, Zed Books Ltd., London, pp 3,4

Grigson, J, (1998) "Jerusalem Artichokes," British food writer in television interview, October 26

Henrikson, R (1997) *Earth Food Spirulina*, Ronore Enterprises, Inc., Kenwood, California, pg. 17

Hobbelink, H (1991) *Biotechnology and the Future of World Agriculture*, Zed Books Ltd., London, pg. 94

Independent UK (2000) "Blair: GM may be a health risk," www.connectotel.com/gmfood/, February 27

Itzkovich, J, (2000) "Diabetes is nectar to the Big Pharm beehive," *The Jerusalem Post,* October 8

Jensen, B and Anderson, M (1990) *Empty Harvest*, Avery Publishing Group Inc., Garden City Park, New York pg. 8

Kaplan, S and Morris, J (2000) "Kids at Risk: Chemicals in the environment come under scrutiny as the number of childhood learning problems soars," *U.S. News & World Report,* June 19, pp 47-53

Klinkenborg, V (2000) "Attracting New Immigrants to the Heartland," *The New York Times* re-printed in *The Jerusalem Post,* September 3

Kaufman M. International Herald Tribune

Leonard, C (2000) "Soybean-Seed Lawsuits Pit Farmers against Biotechnology Companies", *Columbia (Missouri) Tribune,* www.biotech-info.net/soy-seed_lawsuits.html, April 5

Losey, J et al (1999) "Transgenic pollen harms monarch larvae", *Nature,* 399: 214, pg. 2

Nabhan, G (1984) *Enduring Seeds: Native American Agriculture and Wild Plant Conservation,* North Point Press, San Francisco, pg. 97

New York Times, The (1999) "Concern remains as Monsanto drops gene", October 5

NPI Center (1999) "Monsanto's Genetically Modified Milk ruled unsafe by the United Nations ", www.connectotel.com/gmfood/, August 18

Null, G (1990) *Clearer, Cleaner, Safer, Greener,* Villard Books, New York, pg. 111

Oplinger, et al (1999) "Ten reasons why farmers should think twice about growing GE crops", www.uoguelph.ca/www/CRSC/faculty/eac/10reasons.htm, pg. 1

Patent on "Neem" (1999) "Patent on "Neem": Will it deprive Indian Farmers of right to use it?", file: /// C/WINDOWS/ ——-..m/neem products/patentonneem.html, May 4

Peterson, N (1978) "The traditional patterns of subsistence to 1975" Pp 22-35 in V. S. Hetzel and H. J. Frith (eds.) *The Nutrition of Aborigines in Relation to the Ecosystem of Central Australia.* CSIRO, Melbourne, Australia cited in *Saline Agriculture: Salt-tolerant Plants for Developing Countries (1990)*

Report of a Panel of the Board on Science and Technology for International Development, Office of International Affairs, National Research Council, Washington DC, pg. 23

Pfaff, W (2000) "The West's Globalization Drive is Proving a Massive Failure", *International Herald Tribune,* September 29, pg. 6

Powell, B (1986) *South Australian Stock Journal,* vol. 24, pg. 7

Raeburn, pg. (1995) *The Last Harvest,* Simon & Schuster, New York, pp 38,64,121,132-136

Richards, pg. (1985) *The Indigenous Agricultural Revolution,* Hutchinson, London, pg. 25

Rissler, J and Mellon, M (1996) *The Ecological Risks of Engineered Crops,* MIT Press, Cambridge, Massachusetts, pp 49,57-60

Robinson, R (1996) *Return to Resistance: Breeding Crops to Reduce Pesticide Dependence,* agAccess, Davis, California

Rodale Institute (1995) *1995 Project Overviews,* Rodale Institute, Emmaus, Pennsylvania, pg. 25

Sadler, pg. (1994) "Gardening on Ice", *The Growing Edge,* vol. 5, no 2, Winter, pp 36-39

Savage, M (1994) *The Millennial Project: Colonizing the Galaxy in Eight Easy Steps,* Little, Brown & company, Boston, pp 45-48, 510.

Schroder, F (1991) "Plant Plane Hydroponics", *The Growing Edge,* vol. 3, no 1, Fall, pp 62-55

Sears, M and Schaefema, A (1998) "Responsible deployment of Bt corn technology in Ontario", www.cfia-acia.agr.ca/english/plant/pbo/btweb2e.html, pg. 2

Shapiro, R (1999) quoted in *Monsanto Monitor,* introductory issue, January, pg. 7

Shiva, V (1999) *Stolen Harvest,* South End Press, Cambridge, Massachusetts, pp 6-17

Siegel, J (1999) "Robot "farmers" could revolutionize agriculture", *The Jerusalem Post,* July 13

Simpson, B (1967) Native American Story Teller, Yosemite National Park, California, August

Students Against Genetic Engineering (1999) "Genetic Engineering Environmental Risks", www.sage_intel.org/genetic/html, pg. 2

Thomas, L (1991) "Bioponics—Part Two," *The Growing Edge,* vol. 2, no 2, Winter, pp 37-42,65-66

Tudge, C (1988) *Food Crops for the Future,* Basil Blackwood Ltd., UK, pg. 38

Tuxill, J (1999) "Nature's Cornucopia: Our Stake in Plant Diversity", Worldwatch Institute

van der Post, L (1967) *The Bushmen of the Kalahari,* Penguin Books in association with Hogarth *Press,* Great Britain, pg. 31

Vietmeyer, N (1975) Speech at National Academy of Sciences Board on Science and Technology for International Development

Von Liebig, J (1855) *Agricultural Chemistry,* Germany introduction

Wallace, A (1994) *Green Means,* KQED, Inc., San Francisco, pg. 75

Webster's Seventh New Collegiate Dictionary, (1965) G.& C. Merriam Company, Springfield, Massachusetts, pg. 4

Weiss, R (1999) "Seeds of Discord: Monsanto's Gene Police Raise Alarm on Farmer's Rights, Rural Tradition," pAo1 from www.washingtonpost.com/up-srv/WPlate/1999-02/03/1001-020399-idx.html, February 3

Weiss, R (1999) "Farmers fear suicide-seed gene", *The Washington Post* reprinted in *The Jerusalem Post,* February 10

Western Producer (2000) "Triple-resistant canola weeds found in Alta", February 10

White, E and Desnomie, R (1998) "Saskatoon Farmer says Charge Bees and Wind, not Him", *Western Producer,* Saskatoon Newsroom, www.198.169.210.5/current_news/this_week/news/news12.html

Wolfson, R (1998) "Two Courageous Reporters fired for trying to tell story of growth hormone; Monsanto tries to bribe Canadian officials to get hormone approved", www.flipside.org/voll/ar98/bg4-1521.htm, April 15, pg. 3

Yeomans, K (1991) "Allelopathy—Bio-Weapon of the Future", *The Growing Edge,* vol. 2, no 2, Winter, pp 33-36

Appendix 2: Acronyms and Abbreviations

Bt	Bacillus thuringiencis
BOSTID	Board on Science and Technology for International Development
BHT	Butylated hydroxytolune
C	Celsius
CaMV	California Mosaic Virus
CSIRO	Commonwealth Scientific Industrial Research Organization
CSA	Community Supported Agriculture
CGIAR	Consultative Group on International Agricultural Research
CELSS	Controlled Ecological Life Support Systems
DDT	Dichloro-diphenyl-trichloroethane
EPA	Environmental Protection Agency
EU	European Union
GNA	Galanthus nivalis agglutinin/Snowdrop lectin
GM	Genetic Engineering/Genetic Modification
GMO	Genetically Modified Organisms
GNP	Gross National Product
GSI	Global Satellite Imaging
IPM	Integrated Pest Management
LED	Light-emitting diodes
MSG	Monosodium glutamate
NASA	National Aeronautic & Space Administration
NFT	Nutrient Film Technique
OTEC	Ocean Thermal Energy Conversion
PBC	Polychlorinated biphenyl

PKN	Phosphorus, potassium, nitrogen
PVC	Polyvinyl chloride
rBGH	Recombinant Bovine Growth Hormone
RNA	Ribonucleic acid
TPS	Technology Protection System
UNSO	United Nations Sudano-Sahelian Office
USDA	United States Department of Agriculture
US	United States of America
WHO	World Health Organization
WTO	*World Trade Organization*

Abcisse — the separation and fall of fruit, flowers, leaves and other plant parts

Aeroponics — hydroponic system which is based on the circulation of nutrient-rich mist

Agricultural pyramid — the sides of the agricultural pyramid are: plants, farmers, animals, soil

Allele — different versions of the same gene, sometimes in the same organism

Aquaculture systems — aquatic cropping systems to produce plants, mollusks or fish

Arable soil — soil that is suitable for agriculture

Artificial manure — the original name for chemical fertilizer

Bioinnoculants — cultures of bacteria and soil symbionts added to compost

Biometric School — the school of plant genetics dedicated to population breeding

Bioponics — a hydroponic system fertilized organically and inoculated with microorganisms

Biosphere — the earth and all living elements upon it

Biotechnology—-the science of accelerated, transgenetic manipulation of living creatures

Breeding — Controlled propagation of plants to achieve specific purposes

Cash crops — crops which are cultivated to supply raw materials in return for cash payment

Cellulose digesters — modern closed compost bins

Cloches — frames covered with glass or plastic to raise the air temperature around plants

Cold bed agriculture — a system in which plants are nourished with cold water

Companion planting — the planting of different cultivars which benefit from each other's proximity

Compost — broken down, rotted plant material suitable for inclusion in the farm nutrient cycle

Conventional agriculture — agriculture based on the use of fossil fuels, chemical salts and monocrops

Conventional breeding processes — pedigreed breeding resulting in huge stands of identical plants

Crop diversity — multiple cultivars on a farm, multiple varieties within a cultivar

Crop rotation — planting crops is planned succession to curtail insect activity and preserve fertility

Cultivars — cultivated, domesticated plants

Cultivated — raised purposefully by the farmer or agriculturalist, worked by the farmer

Cyanobacteria — blue green algae

Deciduous — loses leaves in the winter

Domestication — the process by which the breeding and propagation of a wild organism comes under human influence

Dormancy — a state of inactivity or suspension of biological processes

Drip irrigation — irrigation systems which deliver precise amounts of water through small pipes

Ecotypical cultivar — a strain of crop plants well adapted to a specific locale

Epistasis — one gene modifies the expression of another gene, not an alelle of the first

Field resistance — enduring, non-specific, polygenetic tolerance of disease or environmental challenge

Gene pirates — nickname for biotechnologists who specialize in finding and patenting genes

Gene transfers — the transfer of genetic material to a different organism's genome

Genetic base — the gene pool of a crop plant

Genetic diversity — the variability and flexibility of a crop's gene pool

Genetic engineering — genetic modification, GM

Genetic imperialism — hunting and using genes from the developing world without giving payment or credit

Genetic modification — manipulation of the genome of an organism by the addition of exogenes

Genome — the complete complement of genes within an organism

Genotypes — two organisms with the same genome are said to have the same genotype

Germinate — to sprout or develop

Germplasm — germ cells and bearers of heredity

Gluten — elastic protein substance in wheat that gives cohesiveness to dough

Gray water — urban but non-industrial waste water, mostly from private dwellings

Growth inhibitors — substances which keep a seed from sprouting

Halophytes — salt loving or salt tolerant plants

Harvest index — the proportion of a crop which can be used for the crop's principle purpose

Heirloom variety — older variations of cultivated plants with specific characteristics, now becoming rare

Hybrids — the progeny of a cross between two different varieties

Hydroponics — literally "water working," growing plants in water

Indigenous — native to an area

Industrial style farming — high input, high energy use factory style farming

Integrated crops — crops and crop successions which are planted together for synergistic benefits

Integrated pest management — limiting pest populations by good farm management

Intercrop — growing two or more cultivars together for their mutual benefit

Isolines — static and uniform end result of pedigreed breeding

Jackpot traits—the "jackpot" traits are: drought tolerance, salt tolerant and nitrogen fixing ability

Landraces — primitive ecotypical cultivars, with great genetic variability and diversity

Legumes — a plant in the pea or bean family, usually capable of fixing nitrogen

Mendelian School — the pedigreed school of plant breeding

Microgardening — producing food in tiny private areas

Modified seeds — seed which has undergone genetic manipulation

Monocrops — crops which are not genetically diverse, frequently the only cultivar in a wide area

Monoculture — the practice of growing large stands of one kind of identical plant

Mucopolysaccharides — the digestible carbohydrates which make up spirulina's cell walls

Mycorrhiza — fungal soil symbiont microorganisms

Neutraceutical crops — GM crops which are grown for medicinal purposes, also called "pharming"

Nutrient Film Technique — hydroponics in which the roots of plants are fed by a thin film of nutrients

Organically grown crops — crops which are raised in healthy soil without chemicals of any kind

Parasitize — to become a parasite upon another living organism

Patented genes — genes which are "owned" by companies, individuals or institutions

Pattern rotation — a sophisticated form of crop rotation which curtails pests and disease

Perennial — a plant which lives for many years

Permaculture — the art/science of creating a stable, sustainable, humane, energy efficient food web

Phyto-pesticides — target materials derived from plants i.e. Neem

Plant clones — identical plants produced asexually

Plant plane hydroponics — simple hydroponic system based a sloping floor and a covered mat of media

Pleiotrophy — effect of a gene on the expression of a number of different genetic traits

Polyunsaturated — oil or fatty acid rich in unsaturated bonds

Propagation — the breeding and multiplication of living organisms

Raceways — oval or rectangular shaped artificial circulating ponds for algae cultivation

Reclamative crop — a crop which is planted to improve impoverished soil

Relic crop — a rare ecotypical cultivar, a surviving "lost crop."

Resistant varieties — plants which demonstrate non-specific or field resistance

Riparian systems — relating to or involving a watershed

Rockwool — fibrous substance used to support roots in certain hydroponics systems

Saponins — any of various glucosides found in plants

Satellite RNA — Small pieces of RNA associated with some strains of plant viruses

Seed Bank — an institution dedicated to the preservation and storage of seeds

Shelf-life — the amount of time a food or product remains edible before spoilage

Single Cell Protein Technology — methods for growing large quantities of SCP

Soil building crops — crops which improve the health and fertility of the soil

Species — a class of individual organisms having common attributes and designated by the same name

Subsistence agriculture — agriculture in which families and villages provide food for themselves

Substantial equivalence — the idea that food plants are not significantly changed by genetic manipulation so do not have to be tested or investigated vis a vis human health

Sustainable agriculture — agriculture which does not damage soil, water or farm environment

Synergistic integration — farm elements which enhance the health and productivity of other farm elements

T-cytoplasm — Texas June male sterility factor in maize

Terminator genes — a GM sequence which causes a plant to poison its own offspring

Tertiary foods — herbs, spices, tea, coffee, cola and chocolate

Toxicity — a state relating to or caused by a toxic substance

Toxin — a poisonous substance which is the product of the metabolic activity of a living organism

Transgene — gene from a dissimilar organism

Transgene traits — traits acquired by transferring genes from another organism

Tubers — thickened underground stems of a plant

V-channel system — hydroponic system which is built around a channel of circulating nutrients

Vermiculite — water absorbing soil amendment used in nurseries

Vertisols — "black cotton" clay soils which are difficult to cultivate

Wildcrop — plant product which is not taken from cultivated plants but gathered from the wild

Wild pathosystems — in which each plant is different so the spread of disease is curtailed by variability

Windbreak — a line of trees or vegetation planted to reduce the force of the wind

Yield drag — the harvest deficit between unmodified crops and GM crops due to GM plants' resource debts

Appendix 4: Characteristics Associated with Weediness

1. Seeds germinate easily in diverse environments
2. Seeds remain viable for many months
3. Plants grow rapidly to the flowering stage
4. Flowers are self compatible so one isolated individual can be the pioneer of a new population
5. Pollination is non specialized
6. Plants produce large numbers of seeds
7. Plants produce seeds in most environmental circumstances
8. Plants are adapted for long distance and short-range dispersal
9. Plants are easily dispersed by weather events
10. Plants regenerate from fragments and have vigorous vegetative reproduction
11. Plants compete by choking growth, producing toxins, forming rosettes or mats of rhizomes
12. Plants resist removal and remain persistently in the soil as regenerating fragments and viable seed

A

B

C

N

O

P

Made in United States
Orlando, FL
04 November 2026

90201125R00219